# American Engravers

C. Roger Bleile

BEINFELD
PUBLISHING, INC.

Published by BEINFELD Publishing, Inc., North Hollywood, California

**Library of Congress Cataloging in Publication Data**

Bleile, C. Roger, 1945—
American engravers.

1. Firearms industry and trade—United States—Biography. 2. Engravers—United States—Biography. I. Title.
TS533.2.B53 739.7'4'0922 [B] 80-18288
ISBN 0-917714-29-6

Designed by Loyal Scott Beinfeld

First Edition

Printed in the United States of America

# Dedication

To those arms engravers of the past, who strove tirelessly to create works of artistic value but passed from this world without adequate recognition of their talent or achievements, this book is respectfully dedicated.

# Contents:

# Foreword

For centuries men have used tools to embellish their possessions with identifying marks or crests. As the art of creative design developed, these tools were also used to embellish and beautify especially prized items. With the increasing demand for such work, the art of engraving began to flower.

Because arms historically have been visual expressions of their possessors' power, they have been decorated, engraved and otherwise transformed from mere artifacts of war into works of art. Over the centuries little has changed. Arms still represent power and they are still frequent objects of decoration. Such embellishment not only enhances their value, it distinguishes individual arms from each other.

The continuing demand for embellished arms has created the need for a book about arms engravers. Most engravers perform their work for a relatively small group of buyers; they rarely have the time or money to promote their work beyond their circle of customers. Because of this lack of dispersion, the collecting public has had no comparative basis for evaluating arms engraving.

Author C. Roger Bleile is a young engraver who was concerned about this limitation. By displaying the work of many engravers, he hopes to provide collectors with the necessary overview of engraving styles and quality.

This book is also a joy to examine. Arms engraving is an art form, as the following pages will demonstrate. Owners of embellished arms can be proud, not only of their possessions, but of their part in furthering that art.

*AMERICAN ENGRAVERS* represents the wishes, dreams and pleasures of many dedicated people, and the gratifying results of their efforts. I hope that you enjoy this work as much as I do.

Wallace Beinfeld

# Acknowledgments

I would like to thank the following persons, who helped make this book possible. First, my thanks to the seventy-three skilled engravers featured in this book, who took time from their work to furnish information and to photograph the arms pictured herein, and who frequently added their encouragement and best wishes.

Second, publisher Wallace Beinfeld believed in and encouraged this project when it was still only an author's idea and freely offered his advice and sound judgment.

Third, Esta V. Bleile ably assisted in the manuscript preparation when the workload overflowed.

Last but not least, I especially want to thank my wife, Catherine, and my daughter Corinne, who bore with me during the months I worked on this project. In addition, Catherine spent countless hours at her typewriter, deciphering my writing and turning out a readable, tidy manuscript.

# Introduction

Because this book is exclusively devoted to engravers and their work, I will start by defining the word "engrave" as it is used herein. *Webster's Dictionary* states that it means "to form by incision (as upon wood or metal)," but most engravers think wood should not be included. To an engraver, his or her skill is applied only to metal, and the incising of other substances would be called carving, sculpting, or scrimshanding.

The art of engraving began at least 5,000 or 6,000 years ago. Examples of the engraver's skill, in the form of seals or signets, have been found in the ruins of the ancient

**Shown here with a floor plate by engraver Lynton McKenzie are his basic engraving tools: (1) chisels, (2) burins, (3) burnisher, (4) scriber, (5) single point matting punch, (6) circle point matting punch, (7) layout compound of beeswax and tallow (McKenzie does not use "white" as do many American engravers), and (8) chasing hammer. The wire at upper left is solid gold and is used for line inlays.**

Babylonian civilization, which flourished as long ago as 3800 B.C. The arms engraver's branch of the trade probably began at the same time, when decorative motifs or symbols were engraved upon prehistoric metallic weapons.

The earliest example of an engraved arm known to me is a bronze dagger in the museum at Athens, Greece. This weapon, attributed to the Minoan culture of ancient Crete, was made and engraved about 1500 B.C., 3,500 years ago. The weapon's bronze blade has been decorated with several flush inlays of gold and silver, depicting men with spears and bows fighting lions.

Today's arms engravers are also skilled carvers and scrimshanders because these skills are required on the stocks, grips, and handles of the arms they decorate; many are skilled gunsmiths and stockmakers as well. Nevertheless, a number of these versatile engravers have told me that they consider engraving to be the most demanding of all their skills.

## VARIOUS TYPES OF ENGRAVERS

In this book the work *engraver* will usually refer to an arms engraver. There are several other types of engravers, such as die engravers, jewelry engravers, bank note engravers, photoengravers, and saddle engravers. These craftsmen are all known as "hand engravers," to distinguish them from "machine engravers," who do lettering and other embellishments with a pantograph or similar mechanical device. It is a sign of our times that machine engravers are becoming more numerous, and hand engravers, more scarce.

To better understand the arms engraver, it will help to know what makes his work different from that of other hand engravers. The die engraver (also known as a die sinker) is a craftsman who cuts steel or brass dies that are used to stamp or mold various items for industry. His handwork is done with the die sinker's chisel, scrapers, and riffler files. Many die engravers also use machinery, such as milling machines or three-dimensional pantographs, to assist in their work. Die engravers almost always execute designs provided by the client. Some past and present arms engravers began their careers as die engravers.

Jewelry engravers almost always engrave soft precious metals and cut with the burin (hand graver). There are relatively few jewelry engravers today because their main forte was engraving monograms and inscriptions, a task which is performed nowadays by pantograph operators, who can be trained in a few hours. The more proficient jewelry engravers are usually capable of executing decorative motifs and figures, as well as monograms and inscriptions. Like some die engravers, a few jewelry engravers have learned the extra skills necessary to become arms engravers.

Bank note engraving is another profession that is nearly extinct in the United States. These craftsmen, working from art supplied by a designer, engrave copper plates which are used to print bank notes or currency. The portraits on American currency illustrate the fine detail of a bank note engraving. Unfortunately for the craftsman, most of today's bank note plates are made by technological methods from existing designs.

Photoengraving is done by a photographic process with the use of acid etching. Companies who specialize in such work for the printing trade employ hand engravers, who touch up printing plates or rollers where the photoetching process failed to properly define an area of the plate or roller. This is extremely tedious work, done by using a burin under very high magnification. Few arms engravers in America have come from this field.

Saddle engravers engrave not saddles, but rather the often elaborate western style accessories and accouterments for horse and riders. Most of the saddle engraver's work is done on silver and is similar in style to that used by jewelry engravers, with zigzag lines and bright cut florals and scrollwork. Silver overlays are used extensively on bits, spurs, belt buckles, and just about anything that the horse and rider can wear or carry in a parade.

I believe a master arms engraver should possess the skills of nearly all the other categories of engravers, *plus* some very difficult-to-master skills not used by any of the others. The true master arms engraver should be able to work in relief, like the die engraver; do lettering, monograms, and light decoration, like the jewelry or saddle engraver; and cut fine line scenes and figures, like the bank note engraver. In addition, unlike any other engraver, the arms engraver usually works on very hard steel. Arms engravers must also know how to inlay precious metals in steel, the most difficult skill to master in arms engraving. As if this were not enough, they must also create the designs for each project.

## HOW ARE ARMS ENGRAVED?

The methods and tools used for engraving and inlaying are fairly simple, but the fine points of the art could fill a separate volume. Of course, it is not my purpose to write a how-to manual on the mechanics of engraving; but a general understanding of how it is done will be helpful to the reader.

Before discussing the tools, I want to stress that an engraving is only as good as its design. Far too many inexperienced engravers start cutting the metal before they have developed their talents for design. I have found that often unattractive engraving is the result of poor design rather than faulty execution.

The reader who aspires to be an arms engraver should study and sketch the best works in this and other books *before* he or she begins to cut metal. Unfortunately, some of today's engravers were either unaware or heedless of this procedure, and as a result their early work shows grave deficiencies of design, as well as execution.

If the ancient engraver Aholiab or the nineteenth century engraver Gustave Young were to visit the workbench of one of our contemporary engravers, he would probably feel right at home with the tools used to execute the designs. The newest tools are special vises and fixtures to hold the work, and devices such as electric polishing wheels used for finishing. Except for these items, the ancient chisel and burin (see Glossary for definitions) are still used by most contemporary arms engravers.

For engraving most modern firearms, cutting and relieving are usually done with a steel chisel and hammer.

Some engravers use the burin, also known as a hand graver or bulino, to cut fine shading lines or bank note style scenes. Because of the toughness of gun steel, few engravers use the burin for outlining scroll or floral work.

Some factory-employed engravers, especially in Europe, receive the parts to be engraved before they are hardened and are able to thus execute scrollwork entirely with the burin. But most American engravers, both free-lance and in factories, have to engrave the gun in its final hardened state. Because of the modern high pressure cartridges, some guns are extremely hard to cut and cause the graver point to continually break or dull. Of course, some gun parts can be annealed (softened by heat) to make them easier to cut, but this must *never* be done with parts subject to stress, such as barrels or cylinders, the very parts that are usually the hardest. Most gun engravers must make their own tools from special steels.

In recent years two devices intended to supplant the hammer-and-chisel method of engraving have been developed. The oldest and best known at this time is a device known as the Gravermeister (German, meaning master engraver); the most recent development is called the NgraveR (see Glossary). Both devices have found favor with some experienced engravers, as well as with beginners. I consider work using either of these tools to be hand engraving since, unlike the pantograph or engraving machine, they are operated completely freehand. Of course, it is important for any beginner to develop skill with the traditional tools, rather than to depend solely on the power-assisted tools.

## THE ENGRAVER, ARTIST OR CRAFTSMAN?

Should arms engravers be considered true artists, or are they highly skilled craftsmen? The issue — which often arises — may never be resolved, but I will offer some observations, admittedly influenced by my own involvement with the subject.

I subscribe to the belief that art is the conscious use of skill, taste, and creative imagination in the production of aesthetic objects. I consider most arms engravers to be artists although, as in any form of art, some are more skillful, tasteful, or imaginative than others.

The owners and admirers of most engraved arms are arms collectors and not "art" collectors. The collector's main concerns seem to be (1) that the engraved piece is more valuable than the unengraved one, and (2) that the engraving makes the subject different from other arms of the same specifications.

As collectors become more aware that the best quality of art and skill is often found on nonfactory-engraved pieces, the demand for free-lance-engraved arms increases. This is especially true for knife collectors, who prize fine handmade pieces. Since the best-made knives in the world have never seen the inside of a factory, they are of course engraved by free-lance practitioners.

This book is intended to enlighten arms collectors about the range of styles and quality that is available today, to familiarize them with the work of the various engravers, and perhaps to help them identify some of their unsigned engravings. Moreover, I am sure that many arms collectors will gain a better appreciation of art in general from their study of engraved guns and edged weapons.

Out of the nearly 200 arms engravers in America, those presented here were not chosen because of my opinion of their skills or artistry, but rather because of their willingness to provide me with the necessary information and illustrations. Their abilities range from acceptable decorators of plain firearms to fantastically gifted craftsmen and artists.

## CONCLUSION

The promotion of a greater awareness of the engraver's art is one of my reasons for compiling this book. I believe that if only a small part of the money used to sponsor other forms of art were spent to encourage new works by engravers, the result would be a dramatic overall improvement of skill and artistry.

More and more gun collectors feel that an important aspect of their collection is to have examples of contemporarily engraved custom guns. The custom knife collectors, much to their credit, seem also to feel this way.

Also in the engraver's favor is the developing trend toward collecting work of specific engravers. In other cases, collectors are seeking pieces engraved by several engravers whose work appeals to them. Such positive trends will have a beneficial effect on arms engraving and collecting, and the art world in general.

It is the true patron who will play a large part in the creation of engraved arms that may be marveled at centuries from now. I hope that, unlike the engraver of the Minoan dagger mentioned earlier, the contemporary engraver will continue to be recognized long after he has passed from the scene.

# Acevedo & Haas, Sr.

**The left- and right-hand Colts inlaid with approximately seven carats of diamonds. The hand-carved ivory grips are inlaid with gold.** (Collection of Buddy Hackett. Photo by Mike Thompson)

These three Colts were originally one of 2,000 Bicentennial sets made by Colt with roll-stamped motifs of a patriotic nature on the cylinder. Mr. Bleile removed the roll stamping and inlaid the cylinder and barrel of each gun with flush bands of gold. The left side of each piece features an inlaid gold motif, while the right side of each has a figure done in bank note style. The custom-made grips are ivory with gold medallions. A set of hand-made knives accompanies this set of Colts; each knife is numbered to coincide with the serial number of the matching gun. (Collection of Brian Accaviatti. Photo by Ed Prentiss)

# Carl B. (Boyd) Bleile

**History buff and Winchester collector Mike Sayers commissioned this Model 12 Winchester to commemorate the 125th anniversary of Wells Fargo in 1977. The shotgun carries a bank note style scene depicting a lawman at the grave of outlaw Sam Bass. The tombstone reads: WELLS FARGO NEVER FORGETS / S. BASS / 1878. This exhibition-grade gun was displayed at the 1977 National Rifle Association national convention and was the subject of a feature length article in the May 1978 issue of** *Shooting Times* **magazine.** (Collection of Mike Sayers. Photo by Dick Pucci)

# C. Roger Bleile

**Engraved by the author in a style reminiscent of that popular on late nineteenth century Colts, this single action army in .45 caliber features antique silver plating, flame-blued screws, and nicely figured ivory grips. Close examination will reveal the engraver's signature, R. Bleile, Eng., on the underside of the ejector rod housing.** (Collection of Richard Penny)

# C. Roger Bleile

Two Ruger No. 1 sporting rifles engraved by the author, Roger Bleile, and custom-stocked by J.N. Hollingsworth of Alvaton, Kentucky. Both rifles are decorated with what the author calls "Belgian leaf scroll." The upper rifle (from the collection of J.H. Pearl) has been finished in French gray. The lower rifle features a bank note-engraved antelope; a bighorn ram's head decorates the side not shown. The engraving has been blued with satin gray finish. (Photo by nationally renowned firearms photographer Charles Twing)

# Acevedo & Haas, Sr.

An exact copy of the famous "Sultan of Turkey" Colt. The original and duplicate are among the most beautiful and graceful of all American firearms. Designed by Mr. Haas, the gun was engraved by Mr. Acevedo. (Collection of Buddy Hackett)

# Acevedo & Haas, Sr.

Among the guns in the original Sam Colt collection is a Gustav Young-engraved and gold-inlaid Third Model Colt dragoon which is considered one of the finest examples of Young's work. Mr. Haas designed and Mr. Acevedo executed this magnificent duplicate. The original is housed at the Connecticut State Library.

# George B. Spring III

Profuse finely shaded full coverage scroll, gold-inlaid cipher, and ivory grips grace this diminutive Bauer stainless steel pistol by Mr. Spring. (Collection of Russ Wehner)

# American Engravers

# Horacio Acevedo

*Kankakee, Illinois*

# & Thomas Haas, Sr.

*Spencer, Indiana*

(Above and right) Details of the duplicate of Colt's Third Model dragoon originally inlaid and engraved by Gustav Young.

In the field of arms decoration and restoration, the combined efforts of Horacio Acevedo and Tom Haas are unique. Their association began in 1956 when Mr. Haas ran a gunsmith shop in Cuernavaca, Mexico during the 1950s. In those days Mr. Acevedo worked in Mexico City, restoring holy vessels for the Catholic church.

Born in Puebla, Mexico, Horacio Acevedo has been a craftsman all of his life. He is not only skilled in the use of burin and chisel, but also does sculpture, carving, and repousse. He has lived in the United States since 1959, when he and Mr. Haas moved to Kankakee, Illinois.

Tom Haas is the designer and gunsmith of the team. He was trained as an engineer and toolmaker during the 1930s. During the 1940s he served as a turret gunner and later as a pilot in the United States Navy. After the war he opened a machine shop, where he did all forms of gunsmithing.

From 1959 until 1964 Mr. Acevedo and Mr. Haas worked together in Illinois. But when Mr. Haas decided to relocate in Spencer, Indiana, Mr. Acevedo opted to remain in Kankakee. In spite of the distance between their homes, the two craftsmen see each other on an almost weekly basis. Their twenty-year collaboration has resulted in the restoration and/or creation of many exceptional exhibition firearms.

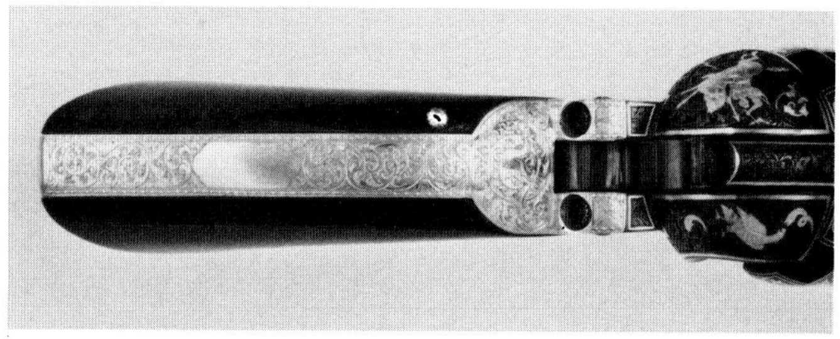

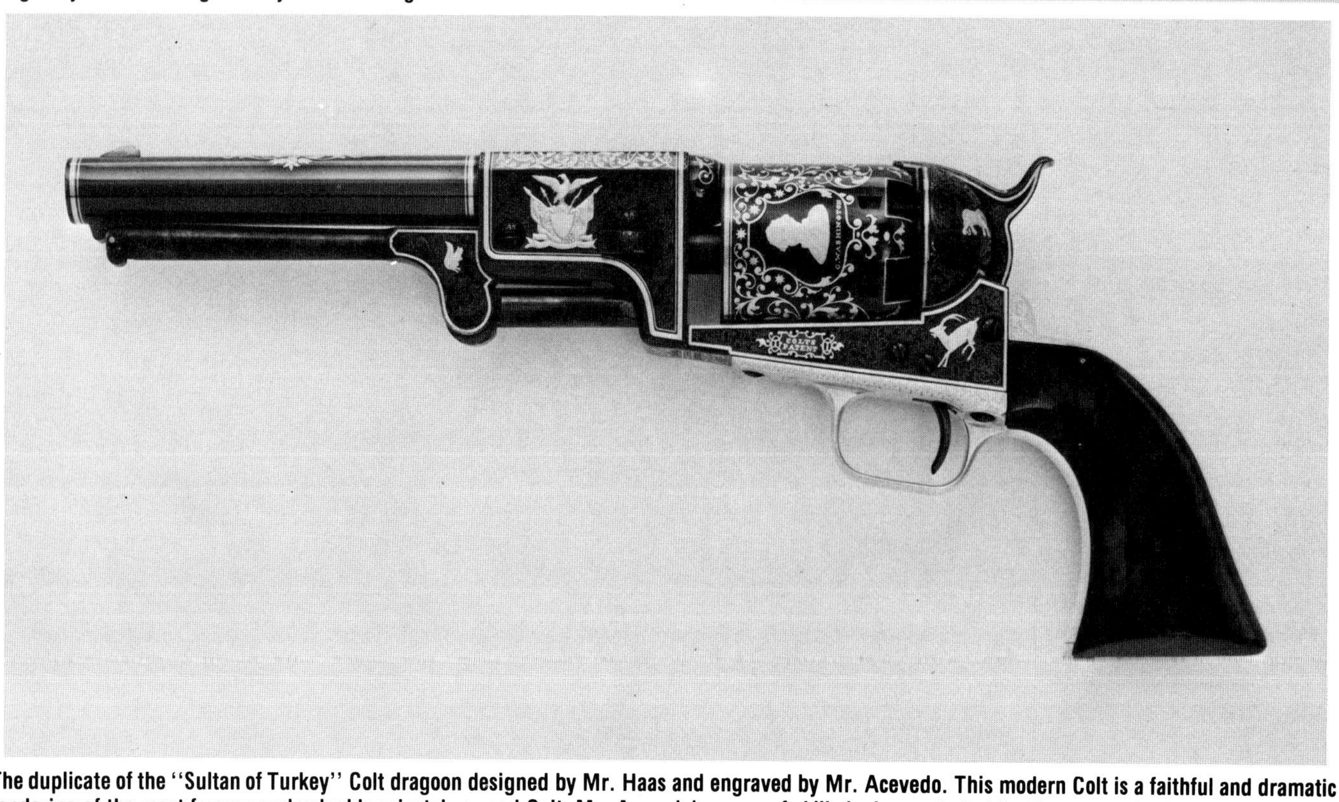

**The duplicate of the "Sultan of Turkey" Colt dragoon designed by Mr. Haas and engraved by Mr. Acevedo. This modern Colt is a faithful and dramatic rendering of the most famous and valuable privately owned Colt. Mr. Acevedo's range of skills is demonstrated by the extremely close engraving and floral and line inlays, as well as the gold lettering and plating. The gun also displays a dozen gold inlays.** (Collection of Buddy Hackett)

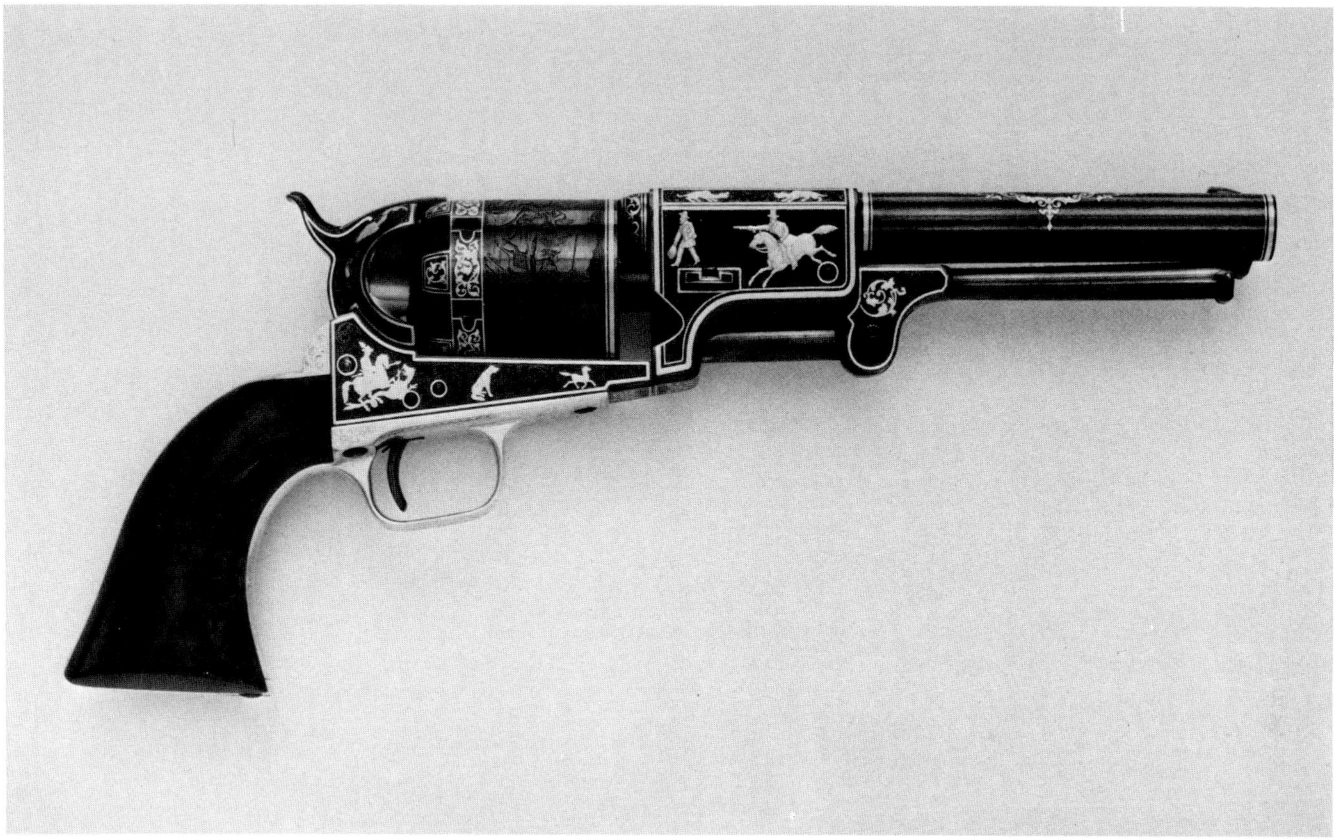

The right side overall view of the Colt Third Model dragoon originally engraved by Gustav Young and exactly duplicated by Mr. Horacio Acevedo under the direction of Tom Haas. The original dragoon is contained in the Samuel Colt collection of the Connecticut State Library. In addition to more than twenty gold inlays, the cylinder, barrel, loading lever, frame and hammer contain fine gold-inlaid bands and scrolls. The cylinder scene is hand-engraved; the backstrap and triggerguard are both engraved and gold plated.

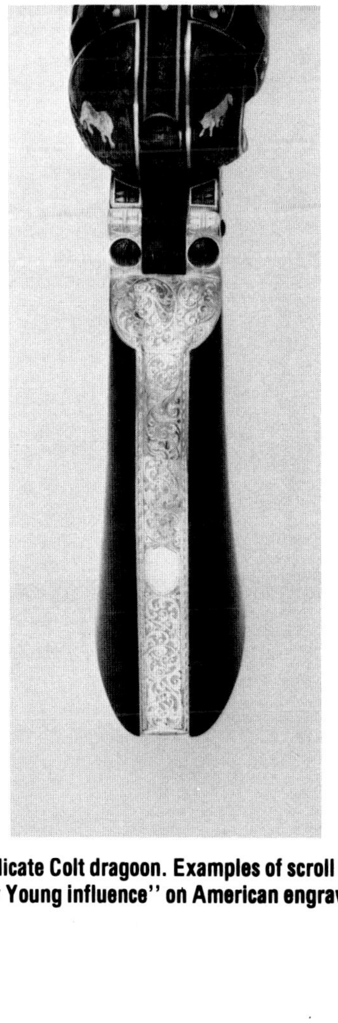

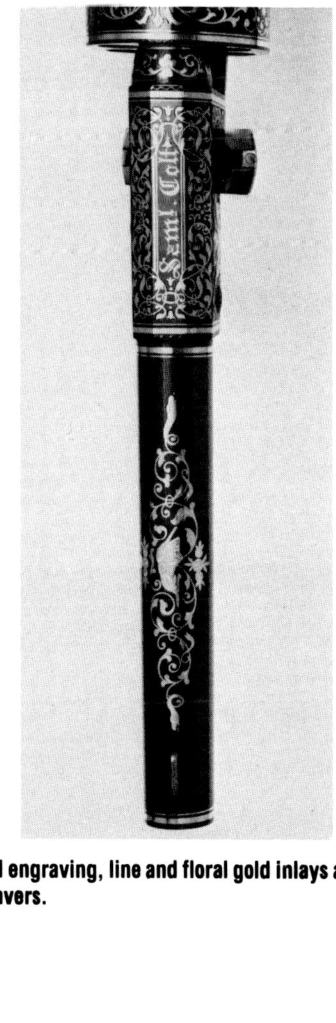

Details of the "Sultan of Turkey" duplicate Colt dragoon. Examples of scroll engraving, line and floral gold inlays as well as animal and letter gold inlays are representations of classic "Gustav Young influence" on American engravers.

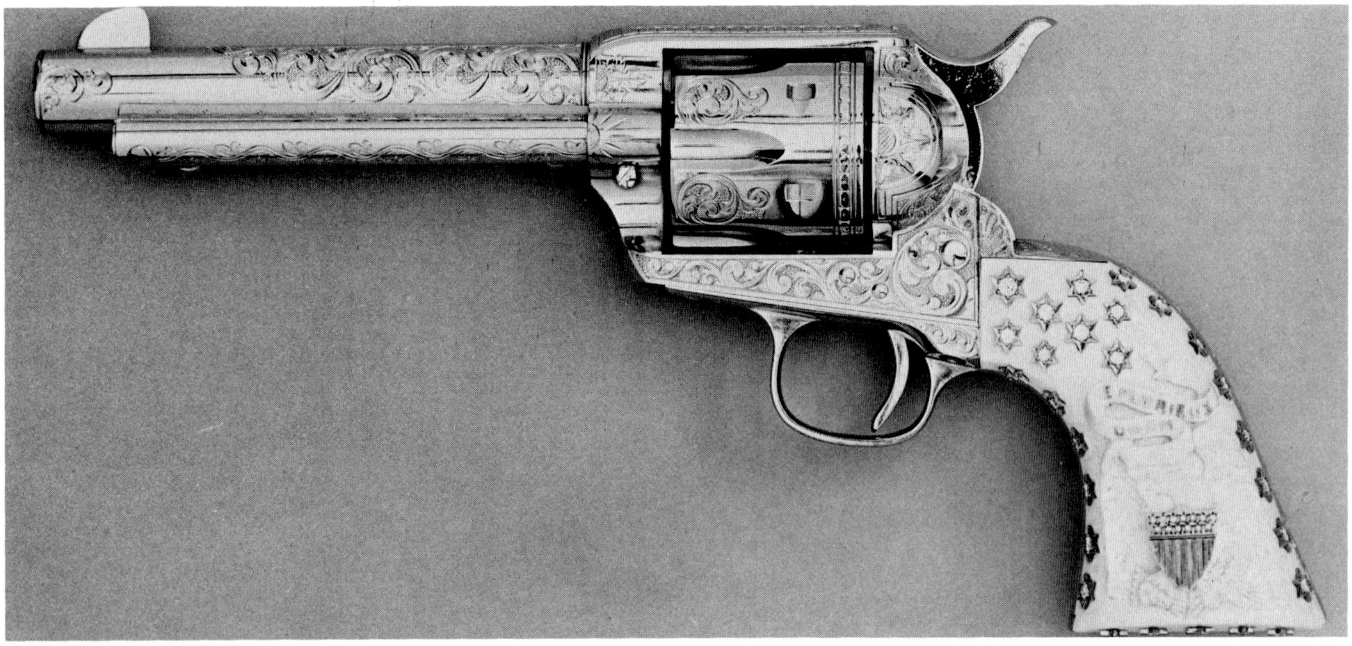

An overall view of the left side of the left-hand Colt.

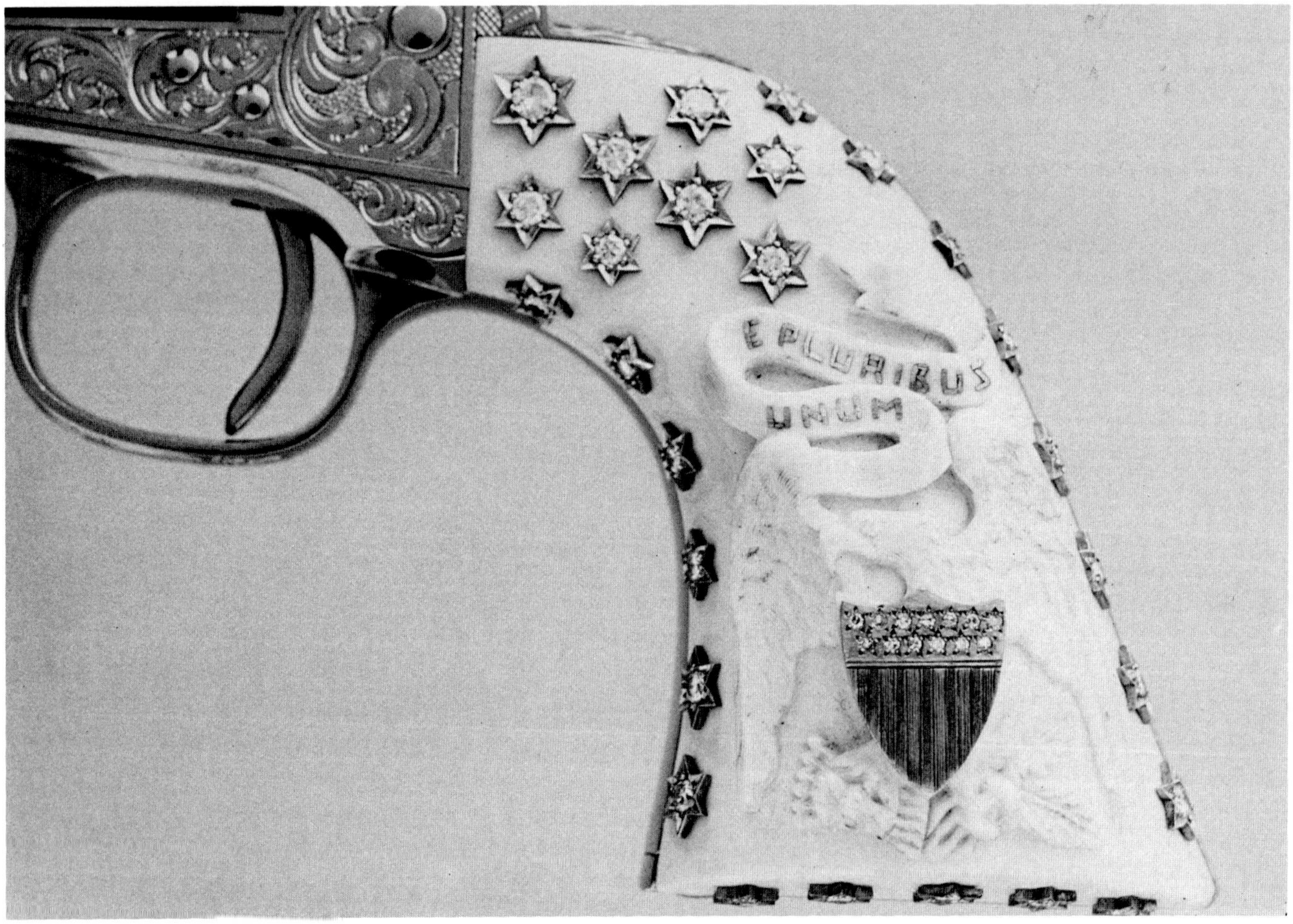


Close-up of the left side of the left-hand Colt. This magnificent gun is a working opposite of the Colt Single Action .45 caliber revolver. Completely engraved, nickel-plated and fully functional, the gun is a mirror of a standard factory model. All the mechanical parts, the triggerguard frame and ejector rod housing were handmade by Mr. Haas. So was the cylinder, which required left turn engaging cylinder stops and reverse ratchet.

Thirteen diamonds representing the thirteen original colonies are set in the gold shield. The elephant ivory grips contain more than 7 carats of diamonds inset in gold. Each star is permanently inlaid in the ivory. The hand-carved American eagle contains the gold-inlaid words "E pluribus unum"; all the minute details are done by hand. The left side of the ivory grips is decorated for the left-hand gun and the right side for the right-hand gun. When worn, therefore, the guns display their highly decorated grips. (Collection of Buddy Hackett)

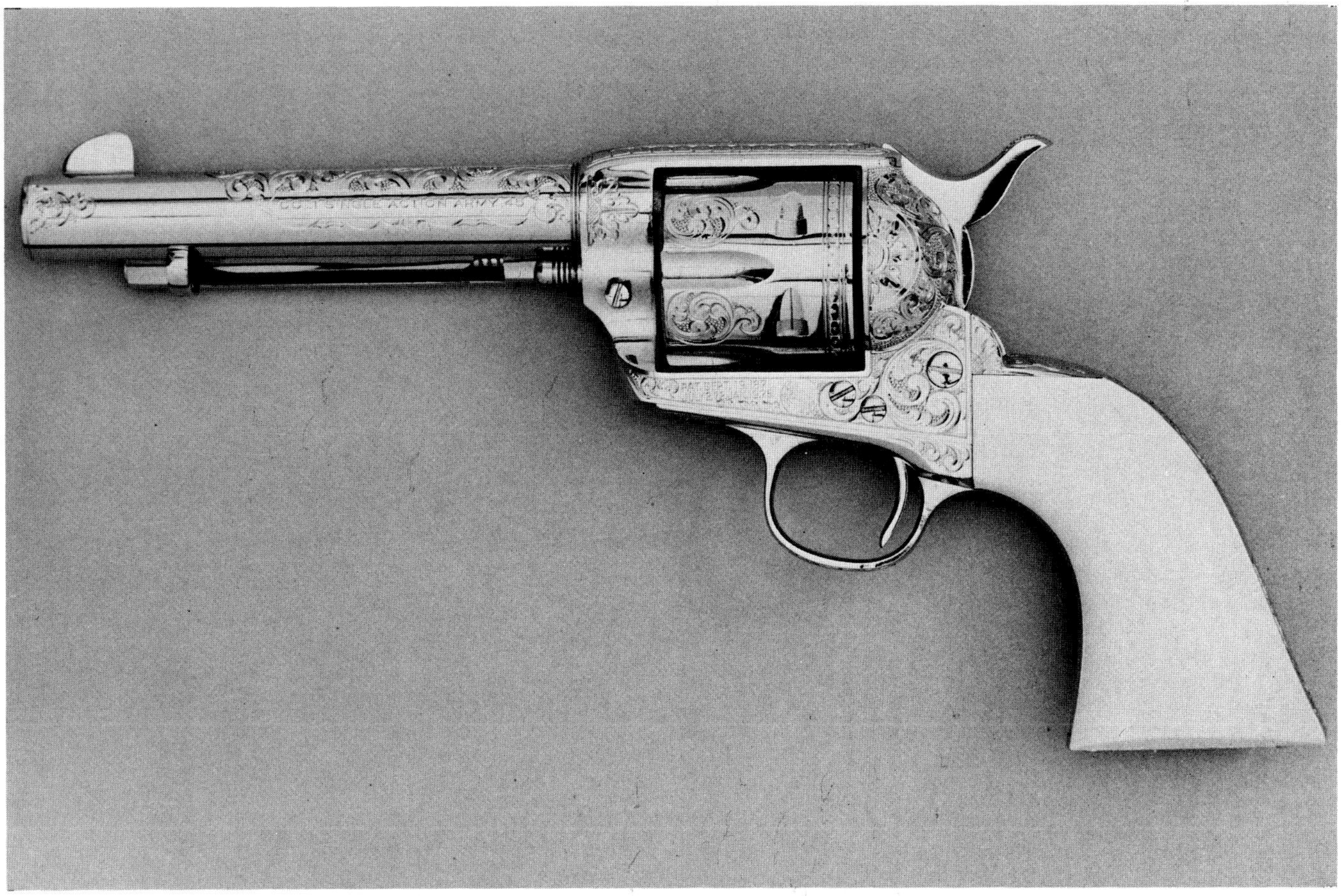

The grips on the right-hand Colt are carved ivory. The outside grip has diamond and gold inlays; the inside grip is plain ivory.

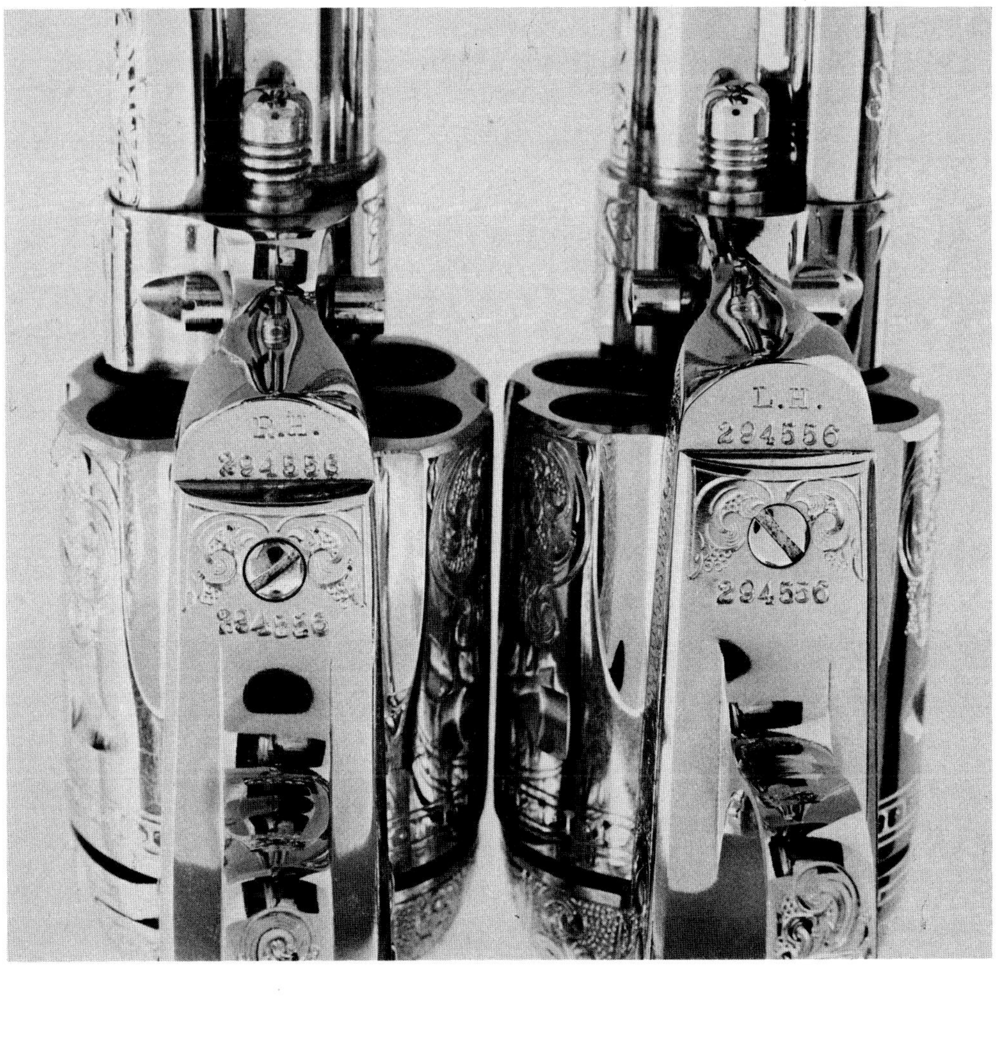


(Left) The frame and triggerguard markings on the Colt factory Single Action revolver. (Right) The left-hand Colt. Note the identifying L.H. markings.

# Paolo Barbetti

*Seattle, Washington*

Paolo Barbetti is a native of Florence, Italy, where he attended the Institute of Art and studied a number of art media, including painting, drawing, sculpture, and metal working. He continued his studies at the Accademia da Arte, also in Florence, where he concentrated on the history of art.

After graduation he became interested in metal engraving, and served an eight-year apprenticeship with Marcello Sacchi, a well-known artist and jeweler in Florence. After the completion of his apprenticeship, he moved on to become one of the primary engravers of coin and metal dies for Mon Arte International, which is an extension of the Franklin Mint.

Mr. Barbetti came to the United States in 1974. He was married in Denver, Colorado, and later moved to Seattle, Washington, where he worked for a short time with Sterling Engraving. When he found that people were very interested in his engraving on firearms, he started his own business, called Paolo Barbetti Artistic Engraving.

For relaxation Mr. Barbetti enjoys tennis, hunting, and trap shooting.

**This is a top view of the Model 62, engraved by Mr. Barbetti. Note the raised, inlaid scrolls on either side of the leaping squirrel and the raised, inlaid border, framing the scroll around the proof mark.**

**Mr. Barbetti has engraved the bottom of this small gauge Parker with German scroll and a line-engraved dog's head with a matted background. The trigger plate has been bordered with inlaid wire. Note the engraver's signature, and the city and date at the top of the trigger plate.**

The scrollwork and figures on this rifle have been executed by Mr. Barbetti in relief, with a raised, inlaid border defining the side of the gun. An oak-leaf and acorn design forms a background for the chipmunk. The rifle is a Winchester, Model 62, in .22 caliber.

The right side of this Parker shows flying quail and scrollwork in light relief.

# Angelo Bee

*Chatsworth, California*

Mr. Bee was born in Italy, but moved with his family to Belgium when he was a small child. At the age of fourteen he began work at the Fabrique Nationale D'Armes de Guerre (F.N.), Browning, factory in Liege, Belgium, where he was to spend his working hours for the next twenty-four years.

At first he worked as a machinist while he also trained as an engraver and studied art at night. According to the system at F.N., engravers began by cutting the relatively simple grade 1 patterns. Next they progressed to the "Pigeon" grade, then to the "Diana," and finally to the "Midas" grade. Only the most experienced engravers worked on the exhibition-grade guns and special projects. Under the eyes of such famous masters as Fernando Funkin, Mr. Bee eventually worked on all grades.

As a result of his many years with F.N., he has engraved over 10,000 firearms, which probably qualifies him for the title of "most prolific engraver in America."

Mr. Bee is presently a free-lance arms engraver, and works for several gun dealers and a few private individuals.

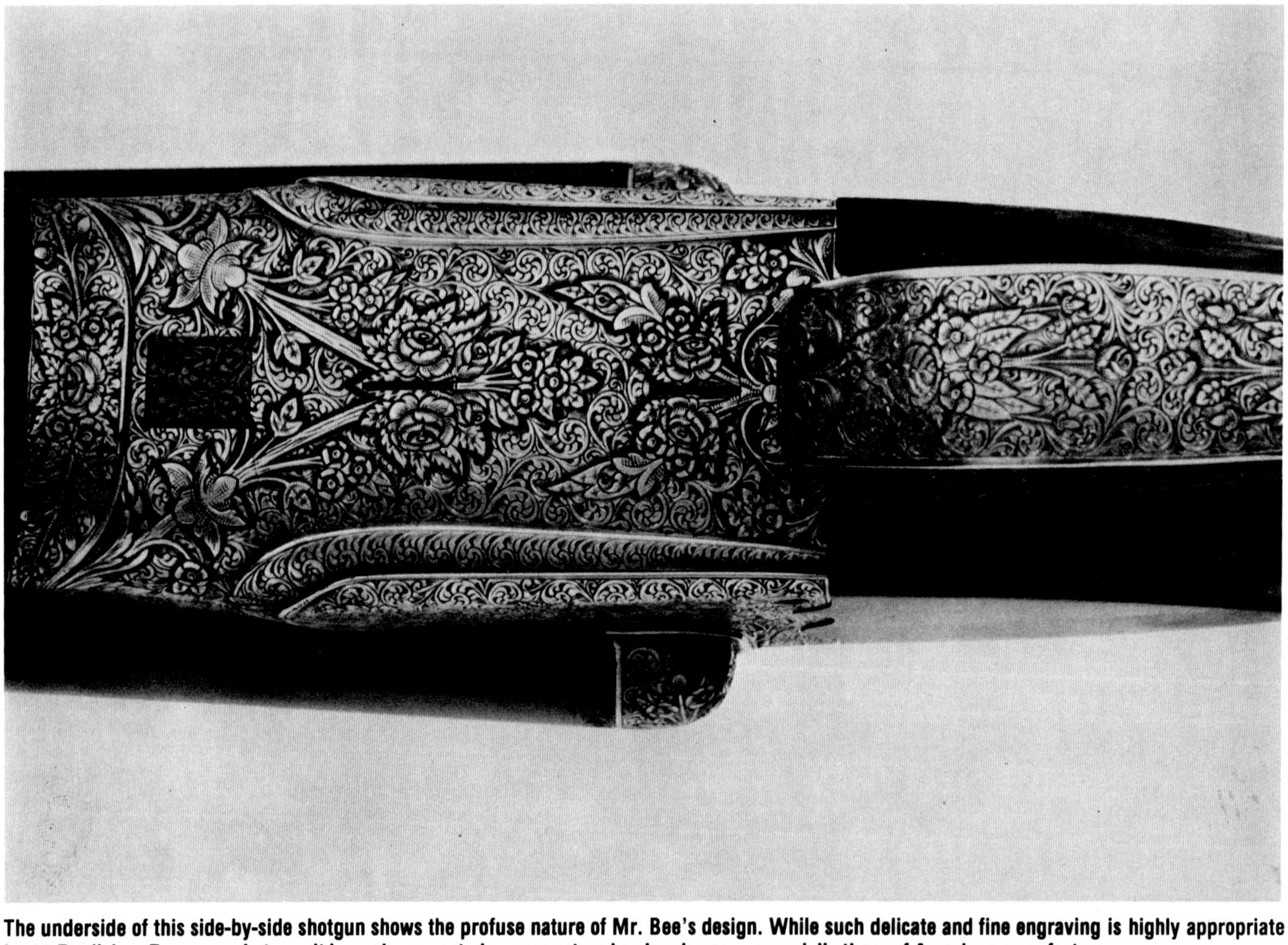

**The underside of this side-by-side shotgun shows the profuse nature of Mr. Bee's design. While such delicate and fine engraving is highly appropriate for an English or European shotgun, it is rarely requested or encountered on handguns — especially those of American manufacture.**

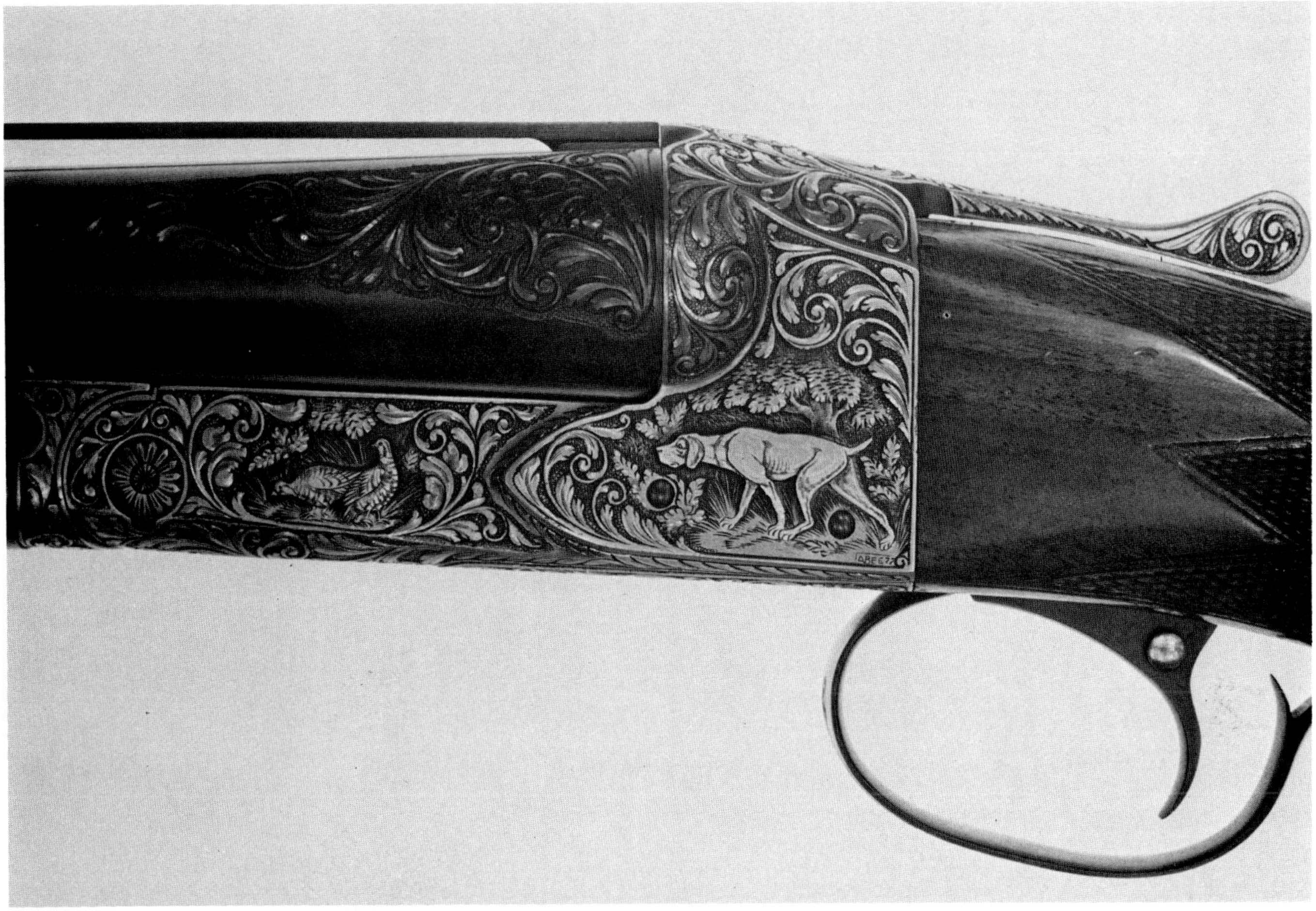

**Medium relief German scroll decorates the frame of this Model 21 Winchester with sculptured relief scrollwork on the barrels, fences, and top of the frame. The inscription ABEE 77 appears on the lower right corner of the frame.** (Photo by Ed Prentiss)

**Made in Belgium, this 12 gauge box lock shotgun was engraved by Angelo Bee, using a variety of well-blended motifs. The scrollwork that serves as a background for the floral designs is a style known as "English" scroll.**

**The extensive inlay work on this gun is even carried to the underside and includes a golden grouse on the trigger plate. The riband on the trigger bow is a traditional motif, intended for the inscription of the owner's name or initials.** (Photo by Ed Prentiss)

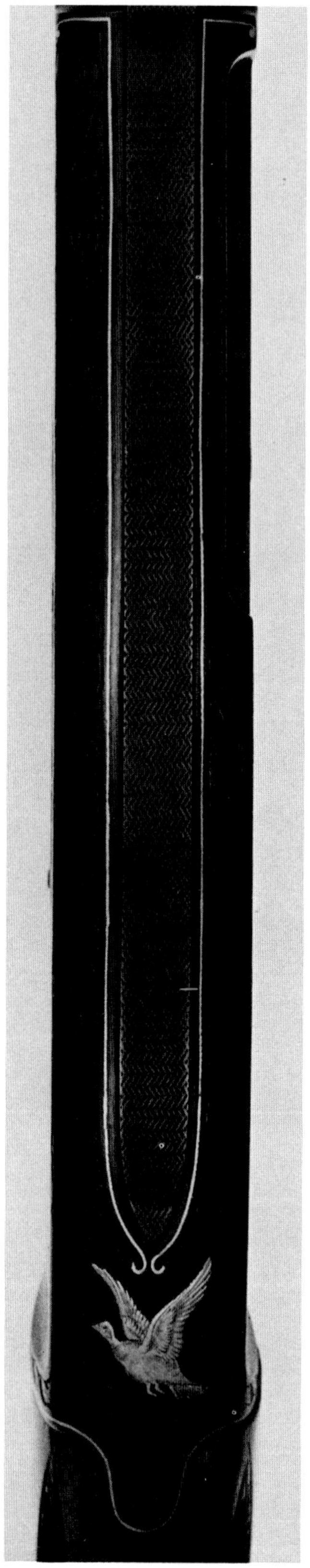

**Top view of the Model 42 shows flush gold borders and a slightly raised gold figure at the rear of the receiver. Note: The sighting rib in the center of the receiver is milled at the Winchester factory to prevent glare to the shooter's eyes.** (Photo by Ed Prentiss)

**Mr. Bee has inlaid the trigger plate of this piece with the classic dog holding a game bird in its mouth and the designation "DE LUXE TRAP" has been inlaid with flush gold wire. Note the gold quail on the trigger bow.** (Photo by Ed Prentiss)

Engraved in a style similar in concept to L. C. Smith's "De Luxe" grade, this side-by-side shotgun sports small and medium scroll with oak-leaf decoration, inlaid in gold. The signature ABEE appears at the bottom of the left barrel where it meets the receiver. (Photo by Ed Prentiss)

Although long out of production, the L. C. Smith pictured here is a good example of a top quality American sidelock shotgun. Though made prior to World War II, this piece was engraved by Mr. Bee in 1977, as attested by his signature below and ahead of the forward-most sideplate screw. (Photo by Ed Prentiss)

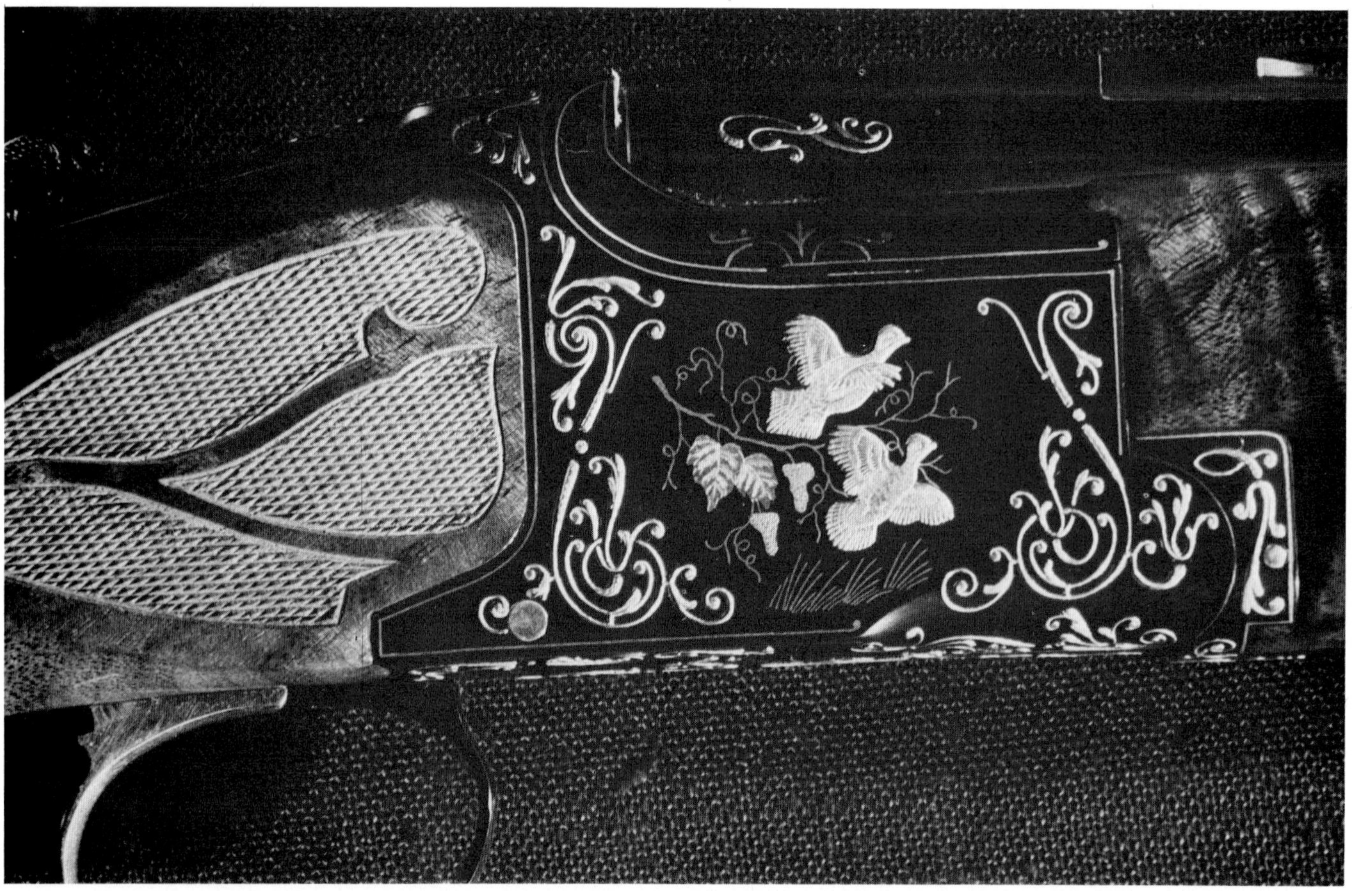

**Flying quail in raised gold are accented by grapes, branches, grass, and grape leaves in flush gold inlay by Mr. Bee.** (Photo courtesy of Bill Jaqua)

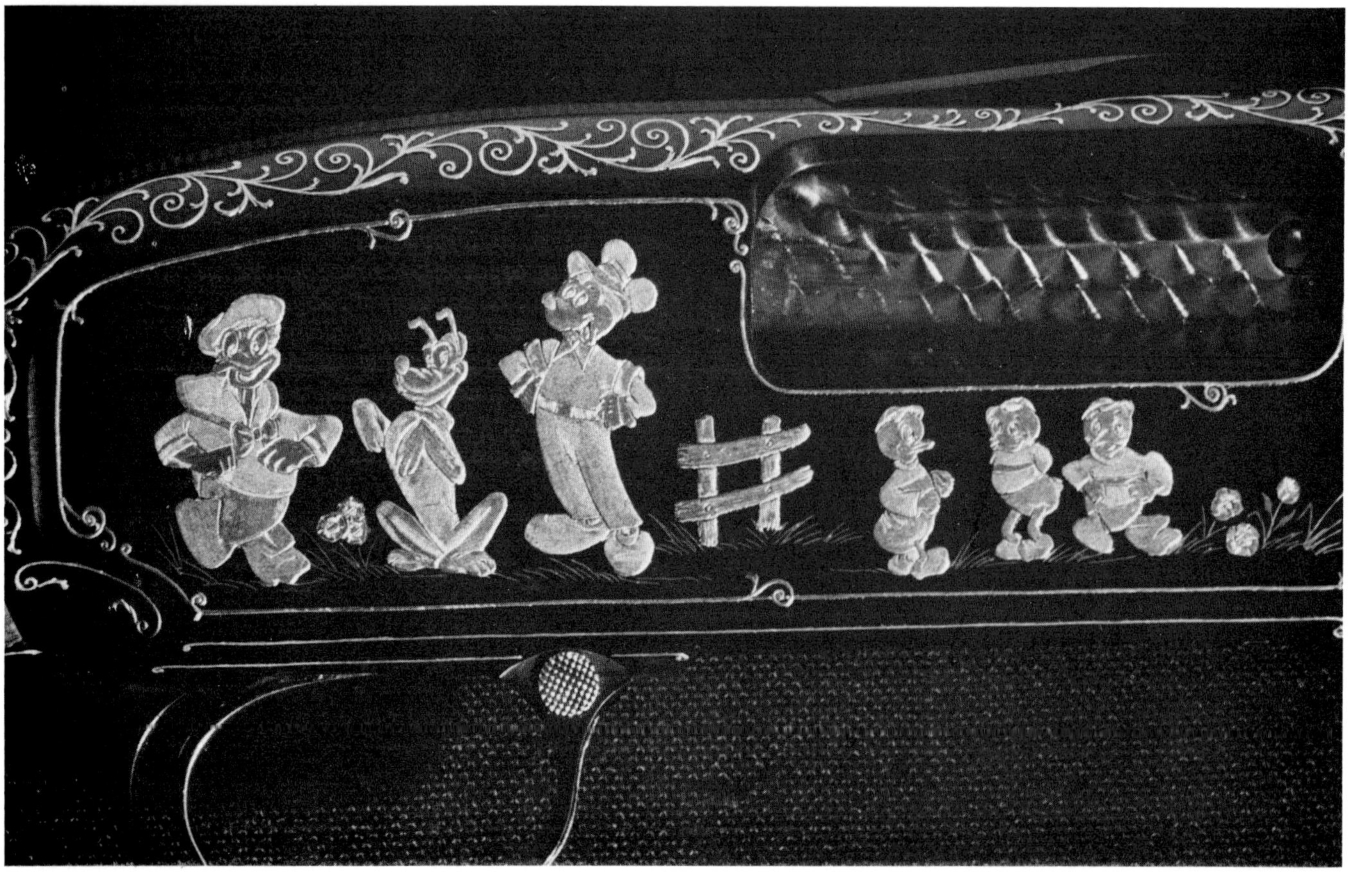

**The right side of Mr. Jaqua's "Disney" gun depicts Mickey Mouse, Pluto, and Donald Duck, with his nephews Huey, Louie, and Dewey. This unusual engraving by Mr. Bee, in a humorous vein, illustrates that all sporting arms need not be decorated with game animals.** (Photo courtesy of Bill Jaqua)

**Inlaid by Mr. Bee for world-renowned sporting arms dealer and collector Bill Jaqua, this unique Winchester Model 12 trap gun is known as the "Disney" gun by trapshooters and collectors. Executed entirely in raised gold inlay, the left side portrays Snow White and the seven dwarfs.** (Photo courtesy of Bill Jaqua)

**Angelo Bee has inlaid this box lock Browning over/under shotgun with gold scrollwork of a slightly geometric style. Note the signature FBEE on the lower right projection of the receiver.** (Photo courtesy of Bill Jaqua)

This Browning Superposed shotgun has been entirely decorated by Mr. Bee in raised gold inlay, with a bit of flush inlay bordering on the barrel and forend attachment. Note the signature ABEE, where the receiver, sideplate, and wood meet above the trigger. (Photo courtesy of Bill Jaqua)

The right side of the Bee-engraved Browning shows a scene of quail surrounded by gold-inlaid scrollwork. This open style of scroll incorporates circular-shaped flowers on the vines. (Photo courtesy of Bill Jaqua)

**A top view of this profusely decorated gun shows the maker's name and grade, inlaid by Mr. Bee into the center rib.** (Photo by Ed Prentiss)

# Bruce K. Beebe

*Akron, New York*

Featured in the *Directory of Native American Artists*, Mr. Beebe is a member of the Turtle clan of the Cayuga American Indian tribe. His memberships also include the Veterans of Foreign Wars and International Veterans Association. He was a sergeant in the United States Marine Corps from 1966 until 1969, the last year being spent on duty in Viet Nam.

Mr. Beebe's experience includes work as a commercial artist, a journeyman layout artist and stone carver, and a teacher of arts and crafts at the Tonawanda Indian Reservation. His scrimshaw work has won awards at the Hartland and Empire State gun shows on several occasions. He is the author of *Scrimshawed Powder Horns, Step by Step Instruction*, published in 1977.

As an arms engraver Mr. Beebe is largely self-taught, but credits engravers Bryson Gwinnell, James DeMunck, and Ray Viramontez, among others, with giving him valuable advice. His choice of engraving tools includes the Gravermeister as well as the chisel and burin.

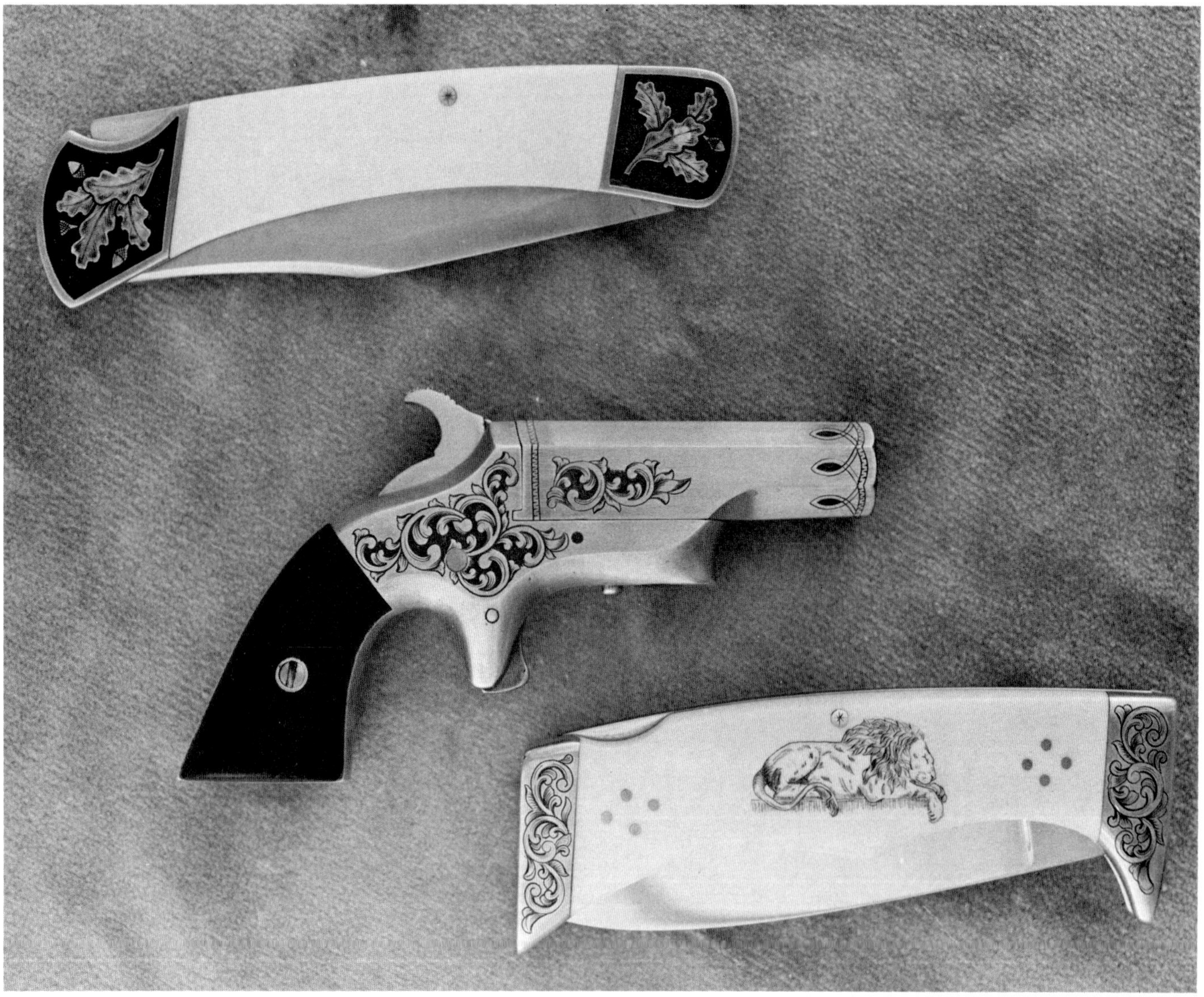

**This selection of arms represents a variety of Mr. Beebe's skills. The Buck folding knife at the top has been carefully fitted with ivory scales and then engraved in high relief with an oak-leaf pattern. The diminutive Southerner derringer is decorated with relief American scrollwork and an intertwining line motif near the muzzle. On the lower knife, Mr. Beebe has called upon his skill as a scrimshander to decorate the ivory scales with a restive lion. The bolsters have been engraved with "American" scroll without background matting or relief. Scroll without background is sometimes called "line work."**
(Photo by Ray Viramontez)

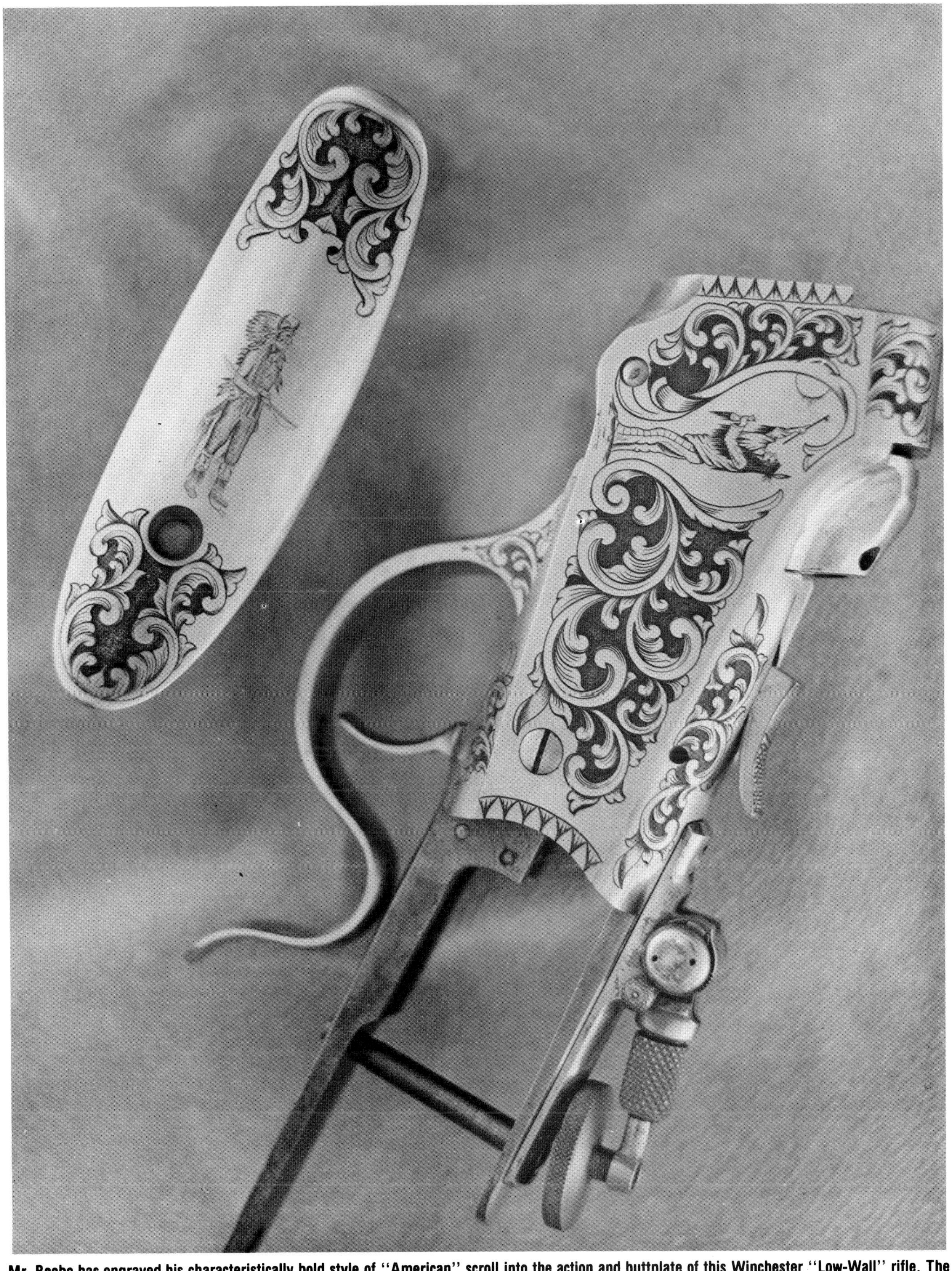

Mr. Beebe has engraved his characteristically bold style of "American" scroll into the action and buttplate of this Winchester "Low-Wall" rifle. The background of the scrollwork has been relieved, matted, and blacked. The American Indian motifs in line cut reflect the engraver's personal heritage. (Photo by Ray Viramontez)

# Carl B. (Boyd) Bleile

*Cincinnati, Ohio*

Carl Bleile (pronounced *Bly-lee*), known to his family and friends as Boyd, is the brother of the author. Like most of his contemporaries, Boyd's interest in both art and firearms goes back to early childhood, when he received his first .22 at the age of six. His initial attempts at pictorial art began at about the same age.

Boyd opened his first business at the age of eighteen. Always seeking a new way to express himself and yet simultaneously earn a living, Boyd became the only tattoo artist in Cincinnati at that time. He also established Cincinnati's health regulations concerning tattoo artists. At the first International Tattoo Artists Convention he was voted "World's Greatest Tattoo Artist" by the other artists, and also vice-president of the organization.

During this period, Boyd also attended Miami University and the Ohio Mechanics Institute of the University of Cincinnati, where he studied physical metallurgy, tool and die making, and mechanical engineering.

After completing his studies, he worked first as a tool and die maker for a number of years, and then as a rotogravure touch-up engraver, continuing his tattooing at nights and on weekends. In 1969 he combined his interests in firearms, art, and engraving, and began teaching himself to engrave guns. Eventually he left his photoengraving job and gave up his tattoo shop to devote more time to arms engraving.

Except for occasional work for Griffin and Howe of New York, Boyd no longer engraves for any dealers or promoters so that he can achieve maximum artistic freedom. Boyd feels that Griffin and Howe, under President Bill Ward, understands his feelings and capabilities. The firm has also employed other fine engravers such as Josef Fugger, Winston G. Churchill, and Robert Swartley.

Early in 1976 Boyd's brother Roger moved back to Cincinnati to complete the study of engraving, which he had begun in California. During the time that followed, Roger Bleile studied and worked with Boyd until December 1977, when he set up his own shop.

Before engraving took so much of his time, Boyd engaged in a variety of interests including karate, sky diving, and car racing. His present diversions include hunting, trap and skeet shooting, painting, stone sculpture, and the casting of precious metal objects. He is a member of the Ohio Gun Collectors Association and the National Rifle Association.

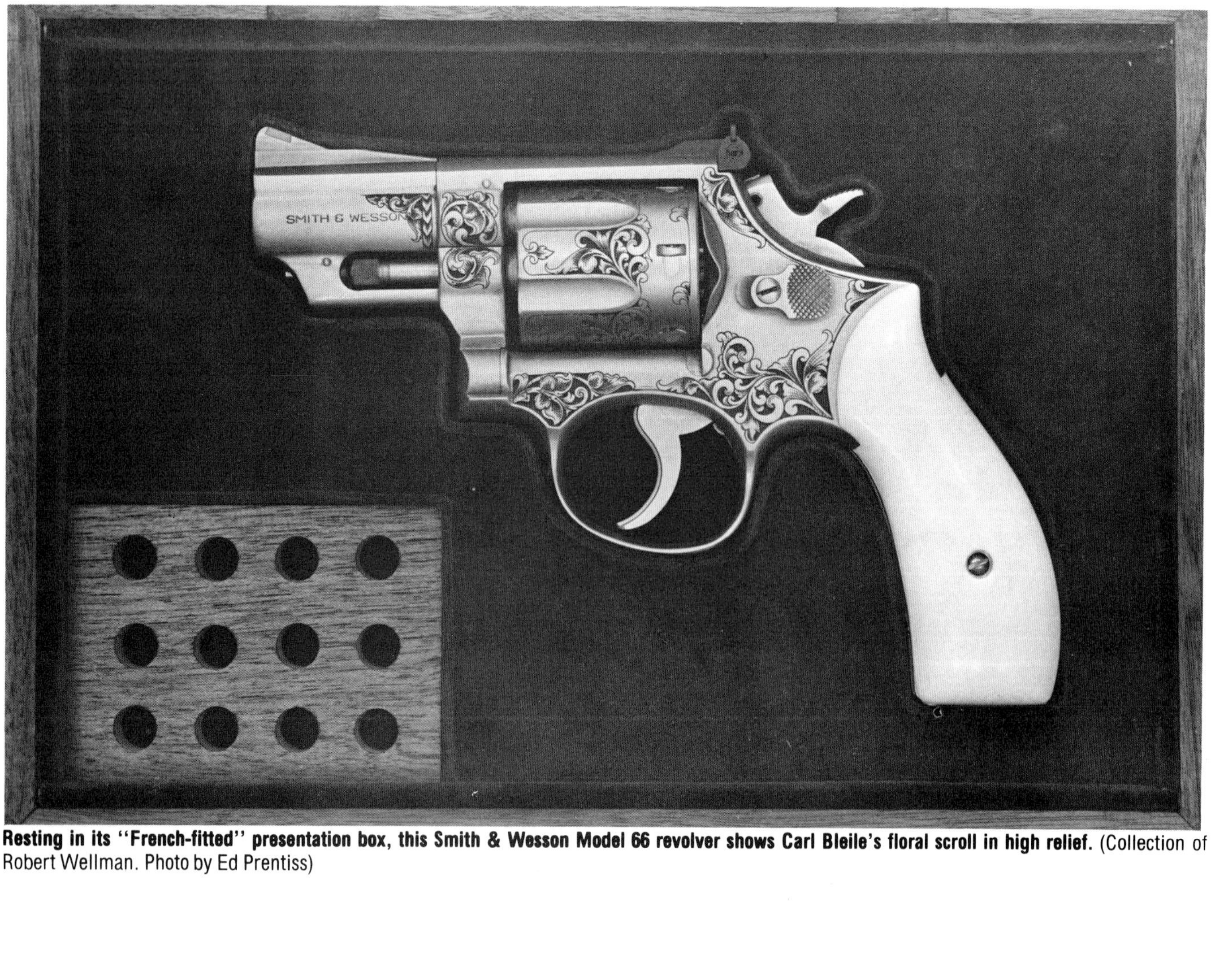


**Resting in its "French-fitted" presentation box, this Smith & Wesson Model 66 revolver shows Carl Bleile's floral scroll in high relief.** (Collection of Robert Wellman. Photo by Ed Prentiss)

**An engraved portrait of Abraham Lincoln was the motif chosen to highlight the relief floral scroll on this Model 66 Smith & Wesson revolver. Note the engraved signature, C. Bleile, Engr., below the bust of Lincoln.** (Photo by Ed Prentiss)

**The saber-toothed tiger scene, magnified here many times, reveals the thousands of individual cut lines necessary to execute a fine bank note engraving.**

**The underside of the Franchi trap gun shows a deep-chiseled relief royal helmet surrounded by high relief floral bordering and a cartouche of tiny Victorian scroll in medium relief. This gun was displayed at the Grand American Trap Shoot by the Stoger Arms Co. and was also featured in the May 1978 issue of** *American Shotgunner.* (Photo by Carl Bleile)

**The reverse side of the "Last Stand" Bowie is adorned with a panoply of American Indian weapons designed and scrimshawed by Mr. Bleile. On the bolster he has engraved a cavalryman and an Indian fighting on horseback; the scene is taken from a painting by Charles Schreyvogel.** (Collection of Bob Nutter. Photo by Ed Prentiss)

Michael Koval of Columbus, Ohio made the Bowie knife shown here. Carl Bleile has sculpted the stag ivory grips with his image of blind justice carrying the all-seeing eye and the sword of enforcement. The relief engraving on the quillons is in a style known as altdeutscher Laubornament or Laubgravur. Many Americans, however, call this style German leaf or floral. (Photo by Ed Prentiss)

This photo shows the bolster of a handmade Lew Booth folding knife magnified many times. The red-tailed hawk has been engraved in the bank note or bulino style. At this magnification, the signature C. Bleile is barely visible below the arch of the tree limb. (Collection of T. Bertolino. Photo by Carl Bleile)

This unusual muzzle-loading percussion pistol was handmade by gunsmith Jack Lewis of Cincinnati, Ohio. Similar guns made by Mr. Lewis have won the National Muzzle-Loading Championship on several occasions. This specimen was engraved during the Bicentennial year and features a bank note-engraved Liberty Bell surrounded by floral scroll decorations in bas-relief. (Collection of Faye Lewis. Photo by C. Roger Bleile)

The receiver of this Ruger No. 1 rifle depicts a panther's head surrounded by a cartouche of high relief scrollwork. The style of this engraving is illustrative of Carl Bleile's early scrollwork and can be contrasted with the style he has used on later works. The panther, though well executed, is much simpler than the highly realistic figures found in his present works. A color picture of this entire rifle can be seen in the 1977 Guns & Ammo Annual. (Photo by Carl Bleile)

# C. Roger Bleile

*Cincinnati, Ohio*

Like most American engravers, Roger Bleile (the author) has been interested in firearms and various forms of art since early childhood, but it was not until his brother Boyd began a career in arms decoration that his own interest in this ancient form of art was sparked.

In 1963 Mr. Bleile joined the United States Air Force, where he served in the Air Police, with duties in law enforcement, nuclear security, and customs inspection. During the Viet Nam war he patroled as a jeep-mounted machinegunner at Tan Son Nhut Air Base. His last two years were spent as a firearms instructor with the rank of staff sergeant.

After military duty he worked as a deputy sheriff in Los Angeles County and then for nearly seven years as a deputy sheriff in Santa Barbara County, California. Among his assignments was that of a shooting instructor for the 250-man department. During his law enforcement career he enjoyed collecting Colt revolvers. He also attended Allan Hancock College, where he received a degree in the administration of justice.

He began engraving by corresponding with his brother and by consulting with engraver Mel Wood (featured later in this book). He then left the Sheriff's Department and returned to his native Cincinnati, where he studied engraving with his brother. Though Roger often attends conventions and collector shows with his brother, their shops have been independent of each other since December 1977.

After carrying a weapon professionally for thirteen years, Roger now prefers to decorate them so that they become objects of art rather than instruments of death. He has written free-lance articles for several law enforcement and arms related publications. He enjoys trap and skeet shooting as well as hunting, and is a member of the Ohio Gun Collectors Association, the Fairfield Sportsman's Association, and is also a life member of the National Rifle Association. His indoor pursuits include visiting museums, and attending concerts of classical music and the opera.

**Engraved by the author in a style reminiscent of that popular on late nineteenth century Colts, this Single Action Army in .45 caliber features antique silver plating, flame-blued screws, and nicely figured ivory grips. Close examination will reveal the engraver's signature, R. Bleile, Engr., on the underside of the ejector rod housing.** (Collection of Richard Penny)

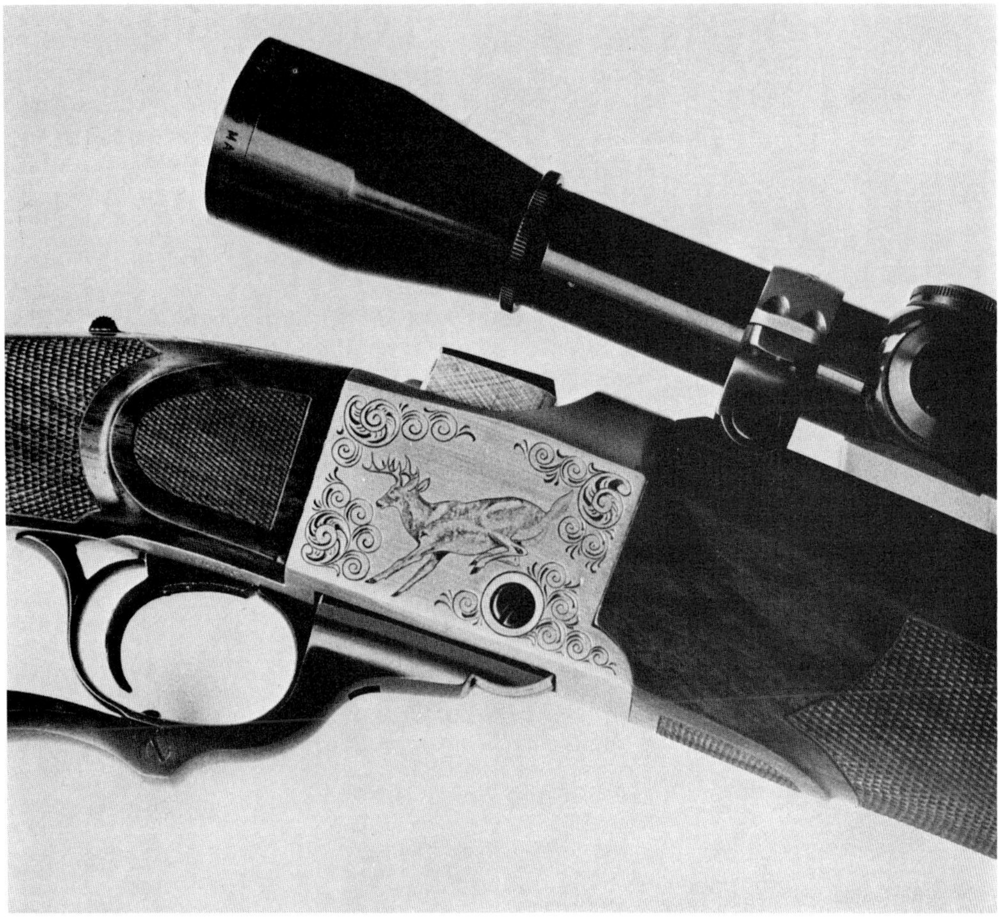

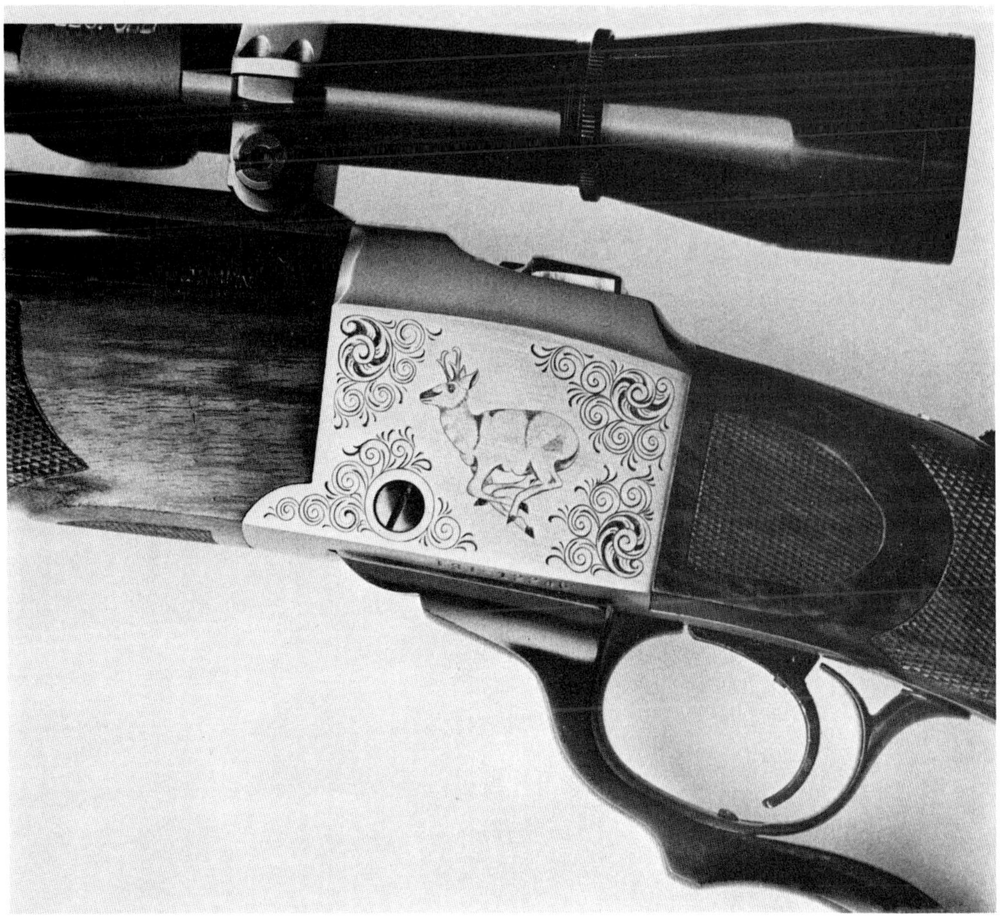

These two views of a single-shot sporting rifle, made by Ruger and custom stocked in Circassian walnut by J. L. Crook, show the author's ability to execute bank note figures. On this piece the author chose to use a style of scroll reminiscent of that used by the late American master John Ulrich. (Collection of J. H. Pearl. Photo by C. Roger Bleile)

# Henry D. (Hank) Bonham

*Seaside Heights, New Jersey*

Mr. Bonham is the former chief of police in a small Maine community and also a former New Jersey state trooper. He is now engraving arms as a full-time occupation. Before free-lancing, he spent a year as an engraver with the Pederson custom division of O.F. Mossburg Firearms Co. There he studied under Ralph Alpen, their chief engraver, who now engraves for the Franklin Mint.

Aside from Mr. Bonham's training in law enforcement at the New Jersey and Ocean County police academies, he majored in art at college, with a minor in psychology. His informal training as an engraver includes guidance from engraver Alvin White and author R.L. Wilson. (Mr. Wilson is the author of *L. D. Nimschke, Firearms Engraver; The Book of Colt Engraving;* and *The Book of Winchester Engraving,* among many others.)

Mr. Bonham's background also includes service during World War II as a Navy signalman aboard the U.S.S. *Bache*, where he saw combat during several Pacific Navy engagements. For relaxation he enjoys north country hunting and fishing, and sings with a barbershop quartet.

After six years as an engraver Mr. Bonham still enjoys his work. He plans to return eventually to Maine, where he likes not only the rustic environment, but also the friendly people.

At present Mr. Bonham engraves for the Champlin Firearms firm in Oklahoma and other sporting goods firms, as well as for private individuals.

**This Mauser rifle floor plate, depicting two deer walking in the snow, is executed in the bank note style. The engraver's signature "HANK" is barely visible on the right end of the plate. Other signatures used by Mr. Bonham include: BONHAM, H.; BONHAM, H.D.; BONHAM; and HDB.** (Boyington collection)

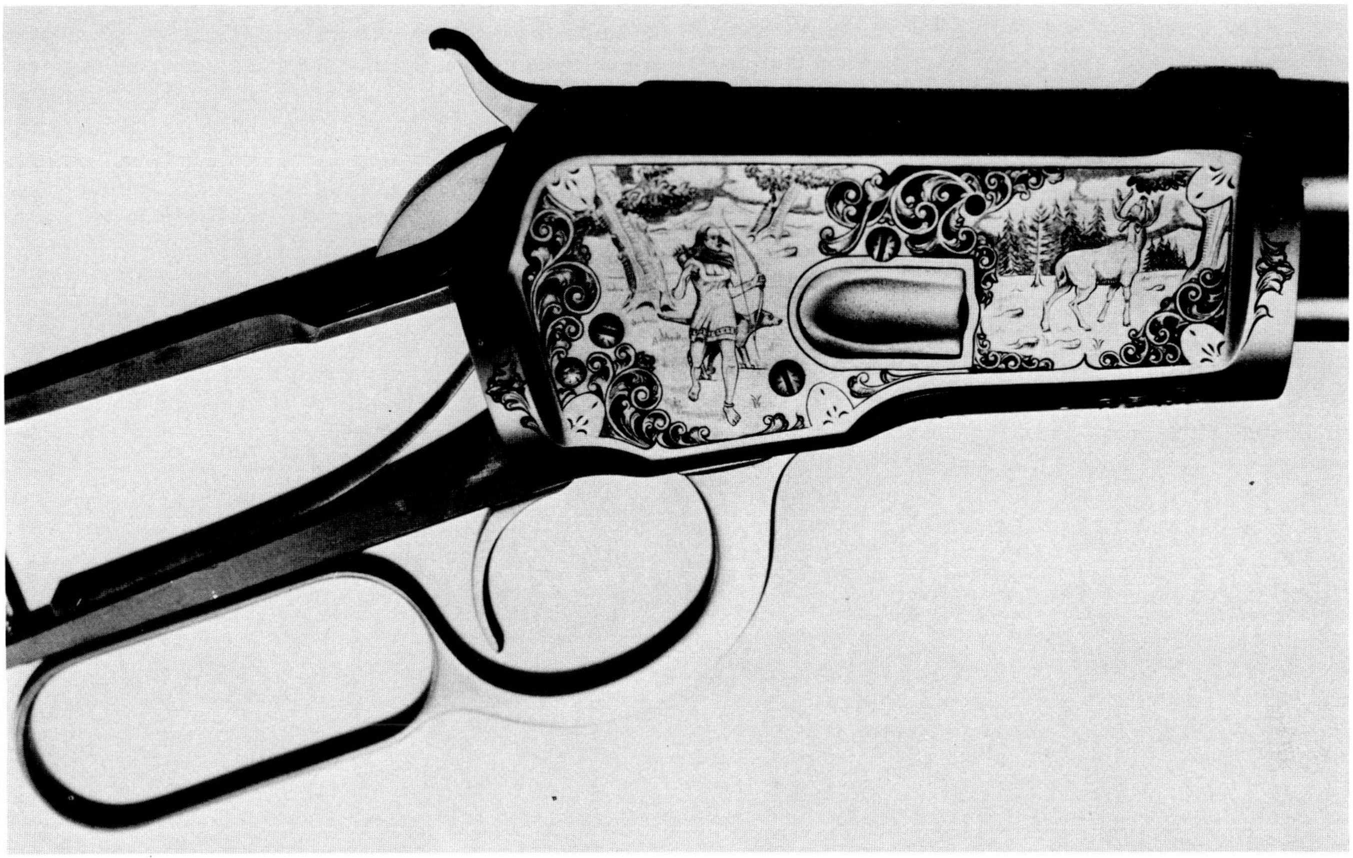

**Engraved with bank note style scenes depicting what Mr. Bonham calls a "Contemporary Age Diana," this Winchester Model '92 is in .44-40 caliber.** (Collection of Willie Dorman)

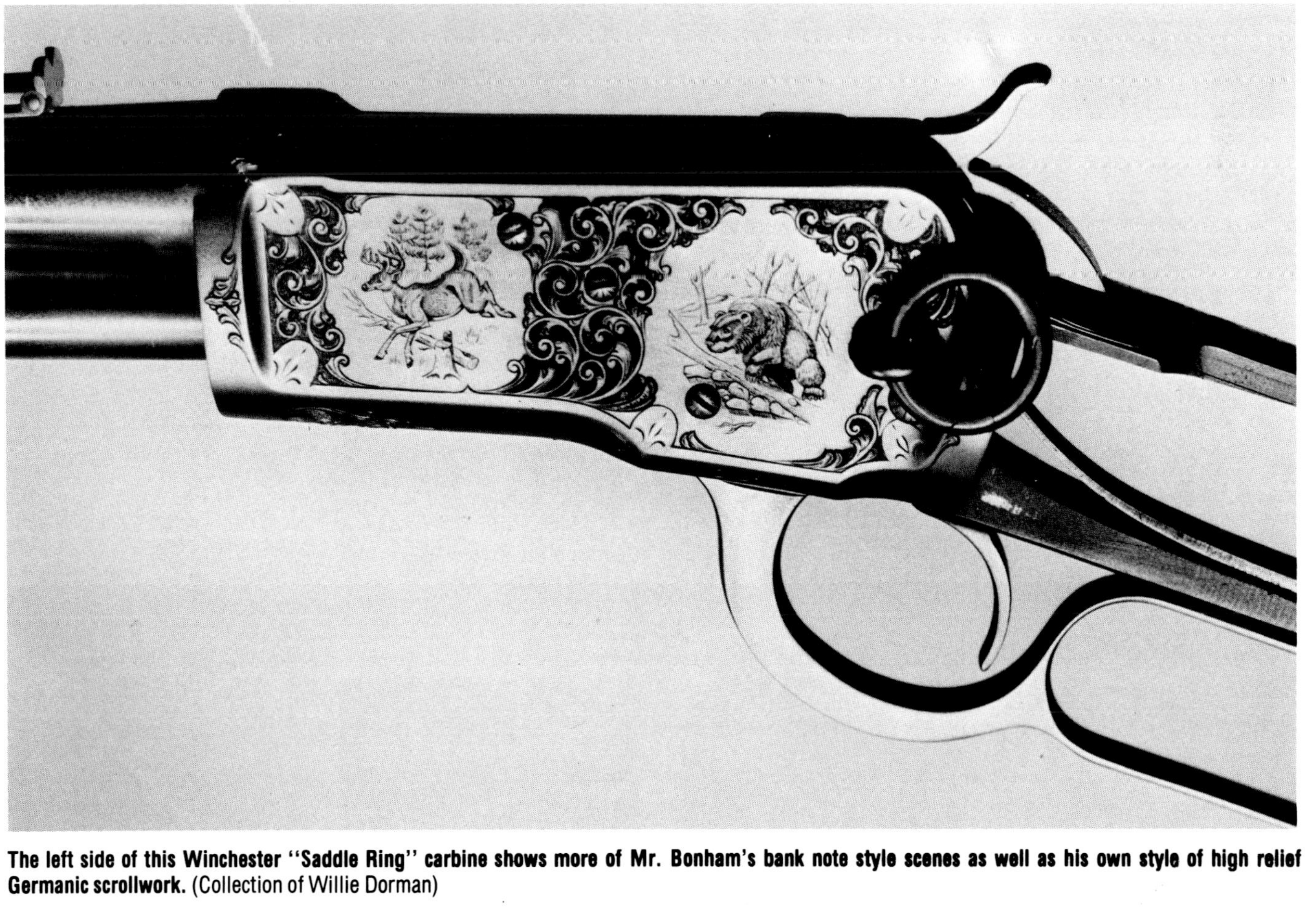

**The left side of this Winchester "Saddle Ring" carbine shows more of Mr. Bonham's bank note style scenes as well as his own style of high relief Germanic scrollwork.** (Collection of Willie Dorman)

# B. Bryan Bridges

*Tucson, Arizona*

Educated at the Chicago Art Institute, Mr. Bridges graduated with honors, receiving a Bachelor of Fine Arts degree. His major subject was sculpture. After college he entered upon a four-year formal apprenticeship in die engraving and received certification as a journeyman die sinker.

After five years with a Chicago firm, Mr. Bridges free-lanced under the business name of Bryan Bridges Sculpture and Engraving. During his career he has hand-engraved dies that were used for a variety of purposes, including paper embossing, gold stamping, coins, medallions, and plastic molds.

In the coin field he has engraved several series of coins and ingots for private mints. Among them were: "Great Patriots," "The Civil War," "America the Beautiful," "American Wildlife," "The World of Flight," "Classic Cars," and the "Postal Commemorative."

Since moving to Tucson, Arizona in 1976, Mr. Bridges has been doing gun engraving, using many of the techniques learned from die work. Most of his tools are handmade; and he engraves by traditional methods, using mostly the chasing hammer and chisel.

Occasionally Mr. Bridges will accept a commission to do a sculpture. He usually starts with clay, which is finished in plaster, and can then be cast in bronze, silver, or gold.

**The receiver of this Perazzi shotgun depicts "Hercules Catching a Wild Horse," inspired by an ancient Greek vase. The scrollwork and birds are in high relief steel, with the central figures in raised-relief gold inlay. Engraved by Mr. Bridges in early 1979.**

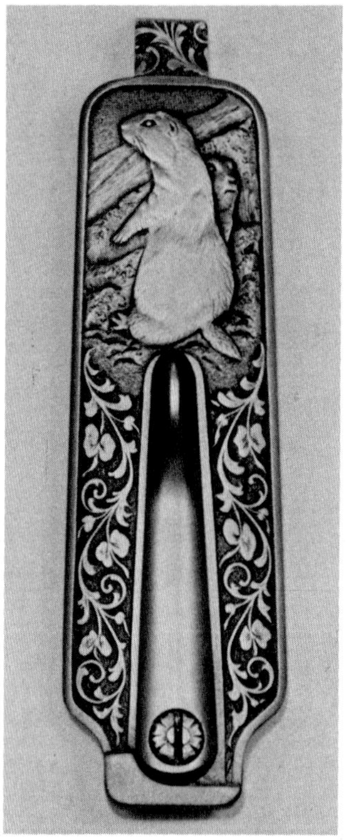

**This Mauser rifle floor plate was engraved in 1977 for the Wilkinson Rifle Ranch, to commemorate their one-thousandth rifle. The prairie dogs and Mr. Bridges's version of German black leaf scroll are in high modeled relief.**

**This grip cap was engraved by Mr. Bridges with high relief scroll and stars.**

**The front and back of this medallion were struck from a steel die engraved by Mr. Bridges for the city of Dallas, Texas in 1976.**

**These wildlife studies are lead proof strikings made from steel dies, hand-engraved by Mr. Bridges for the Hamilton Mint.**

Engraved in 1978, the receiver of this Krieghoff over/under shotgun illustrates Mr. Bridges's flare for the unusual. The Hopi Indian dancers are inlaid in raised-relief gold. The background is also in high relief and is gray finished.

This floor plate of a bolt action rifle portrays a lion engraved in deep modeled relief and surrounded by what Mr. Bridges refers to as "African scroll."

The figures on the underside of this Krieghoff are relief-engraved and modeled from the gun steel. Matting with the circle punch is used for the background of the figures as well as for the high relief scroll-work.

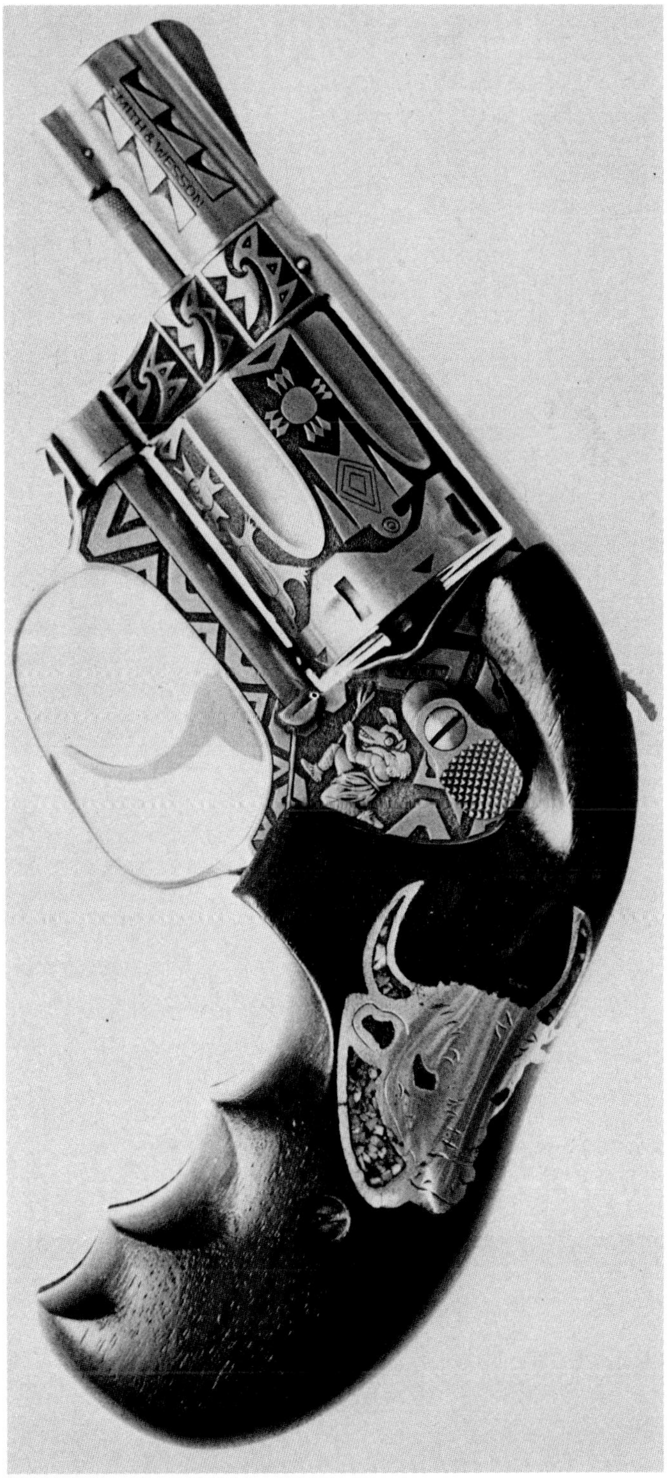


This unique stainless steel Smith & Wesson revolver has been engraved by Bryan Bridges with Southwestern Indian designs and a Hopi dancer inlaid in gold. The grips are inlaid with silver and turquoise.

# Gino Cargnel

*Chatsworth, California*

A recent arrival to the United States, Gino Cargnel is a native of the famous arms-making city of Liege in Belgium. There he worked at the Fabrique Nationale D'Armes de Guerre (F.N.), Browning, factory as an engraver from 1954 until 1978, when he emigrated to the United States.

During his twenty-four years at F.N. he worked side by side with Angelo Bee (featured earlier), and they learned arms engraving from the same masters. The two men were also friends outside the walls of the famous gun factory. They spent some of their free time using their construction skills to build home additions in Liege.

Gino is married and has two sons. His hobbies are fishing and oil painting.

**Mr. Cargnel has decorated this Winchester Model 12 in gold with nearly every motif suggestive of the city of Las Vegas. Las Vegas is well-known among gun enthusiasts for the twice-yearly gun shows and fine trap shoots.** (Collection of Bill Jaqua)

**A cowboy roping a wild horse decorates the receiver of this Krieghoff Model 32 over/under shotgun. The figures are of raised inlaid gold. The bas-relief scrollwork is a large, flowing type often used by Mr. Cargnel.** (Photo courtesy of Bill Jaqua)

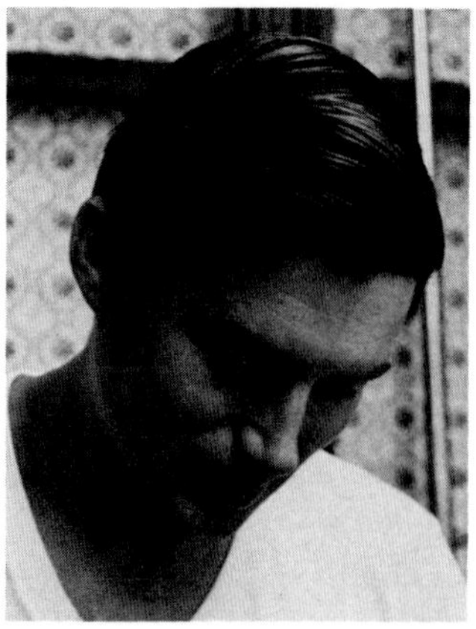

# Winston G. Churchill

*Proctorsville, Vermont*

By making his own toys from wood at an early age, Winston G. Churchill began laying the groundwork for a life as a craftsman and artist. At the age of thirteen, he made his first gunstock for an "ancient" shotgun that was a gift from his grandfather, whittling the stock from a white ash plank.

In high school he took a cooperative course in industrial pattern making, becoming a journeyman a year after graduation. It was during his high school years that he was influenced by another pattern maker, Dave Farr, who made gunstocks in his spare time. Mr. Farr taught young Churchill the fine points of guns and wood and guided his progress in the stockmaker's craft.

Entering the Navy in the late 1950s, Mr. Churchill was assigned to the pattern shop of the U.S.S. *Vulcan*, a repair ship. One of his duties was to relief carve ships' insignias on mahogany plaques. These plaques were presented to various dignitaries by ships' captains. Thus his carvings were distributed around the world.

During his tour of duty he met and married his wife, Fran. After leaving the Navy, Mr. Churchill worked at various jobs in Vermont until, in 1969, he was offered employment with Griffin and Howe in New York. Although he had taught himself many of the basics of engraving, his apprenticeship at Griffin and Howe under the direction of Josef Fugger (pronounced Foo'ger) polished and refined his skills. Mr. Churchill maintains that the now retired Fugger was the best engraver of his generation, a belief that is widely held.

At Griffin and Howe (then owned by Abercrombie & Fitch) Mr. Churchill also checkered gunstocks while studying and practicing engraving. He worked with Mr. Fugger for three years until the retirement of the Austrian master and remained at Griffin and Howe for the next year and a half. He then returned to Vermont to do free-lance engraving.

Since his return he has built a home on the farm where seven generations of the Churchill family have lived since the Revolutionary War. Most of his engraving is performed on fine firearms, but he also occasionally engraves fine handmade knives and a very few pieces of jewelry. A large part of his time is spent engraving cased exhibition pieces.

Fine detail is Mr. Churchill's specialty. He does only museum quality, "best grade" work, executed in several styles. Each design is carried out using only traditional hand tools, and all cutting is done with a single point graver.

Mr. Churchill has no apprentice or employees, and has no plans to hire an assistant. But his two sons are ten and twelve, and he feels they show artistic aptitude. Perhaps he will be fortunate enough to pass on his knowledge and skills to his sons.

The right side of this Smith & Wesson reveals that the factory stampings on the sideplate and forward part of the frame have been completely removed to provide an unbroken area for the flow of the design. Note the highly detailed rosette treatment given to the sideplate screws.

This rifle sight has received an Arabesque treatment by Mr. Churchill. Note the beautifully rosetted hinge pin.

Detail of the buttstrap illustrates the gun's serial number inlaid with flush gold and the highly detailed style of Germanic scrollwork for which Mr. Churchill is best known.

The exhibition-grade Smith & Wesson pictured in this and the accompanying photographs is noteworthy for the crispness of its metal work. It was completely hand-fitted, filed, and polished prior to any engraving or inlaying because Mr. Churchill feels that proper preparation is essential to any fine exhibition-grade arm. In order to reveal the fine detail work, all black and white photos of this gun were taken prior to bluing.

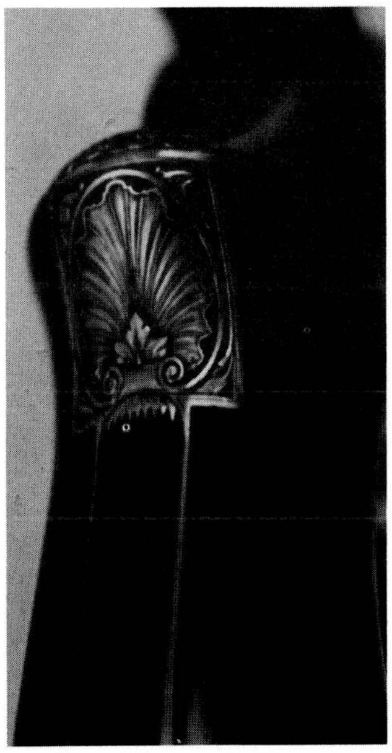

Detail of the grip tang shows more of Mr. Churchill's talent at sculpted relief steel, here in the form of a scallop shell.

Detail of the forward part of the frame below the barrel. Note that the joint between the crane and the frame is barely visible.

This Model 32 Remington was stocked, inlaid, and engraved by Winston G. Churchill. The designs include Germanic scroll, as well as what Mr. Churchill calls "Arabesque" scroll. The scene on the right of the receiver depicts a Brittany spaniel flushing three grouse. All are in relief gold.

What is not apparent in this photo is the gold inlaid inscription: MODEL 32/5403/SKEET. The inscription is engraved in what looks like a blank space below the Remington name. Also barely visible is the engraved signature, WINSTON CHURCHILL, ENG., in block letters alongside the trigger.

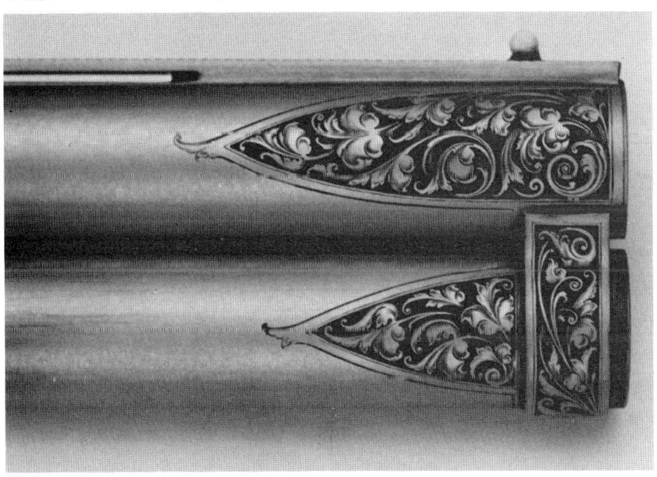

Mr. Churchill has used this Model 32 Remington shotgun as a vehicle for his artistic expression. He has combined German leaf ornamentation (which he calls "Arabesque") on the barrel and top slide with what the Austrian school calls neuenglische Arabesken on the receiver. The animals in the game scene are relief sculpted in gold.

Detail of muzzles shows "Arabesque" design, known by some engravers as black leaf or floral.

Enlarged detail of the grip cap shows minute gold inlaid quail, actually less than 1/4 inch long.

Details of the revolver's two cylinders. The central figure on each flat is inlaid and sculpted gold. Note the beautiful polish of the flutes.

Detail of the right side of a Remington rifle engraved and stocked by Winston Churchill. The photo shows the scope mount base and rings. Note the ciphered initials DEP on the scope mount.

Detail of the rifle's grip cap. Note the wheat chaff border and rosetted screw head.

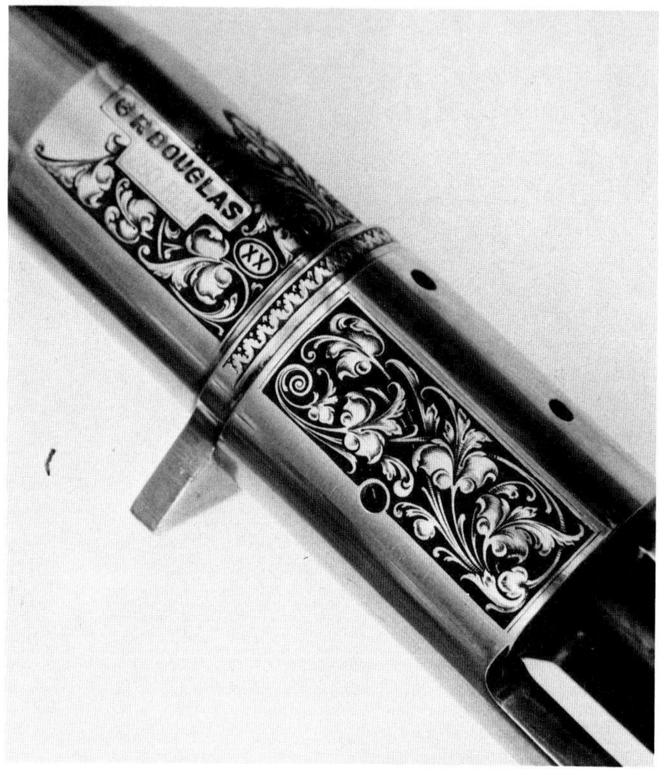


Detail of the rifle's barrel and action. Note the fine single point shading of the high relief floral designs.

This gold ladies' brooch illustrates Mr. Churchill's work on a nonarms subject.

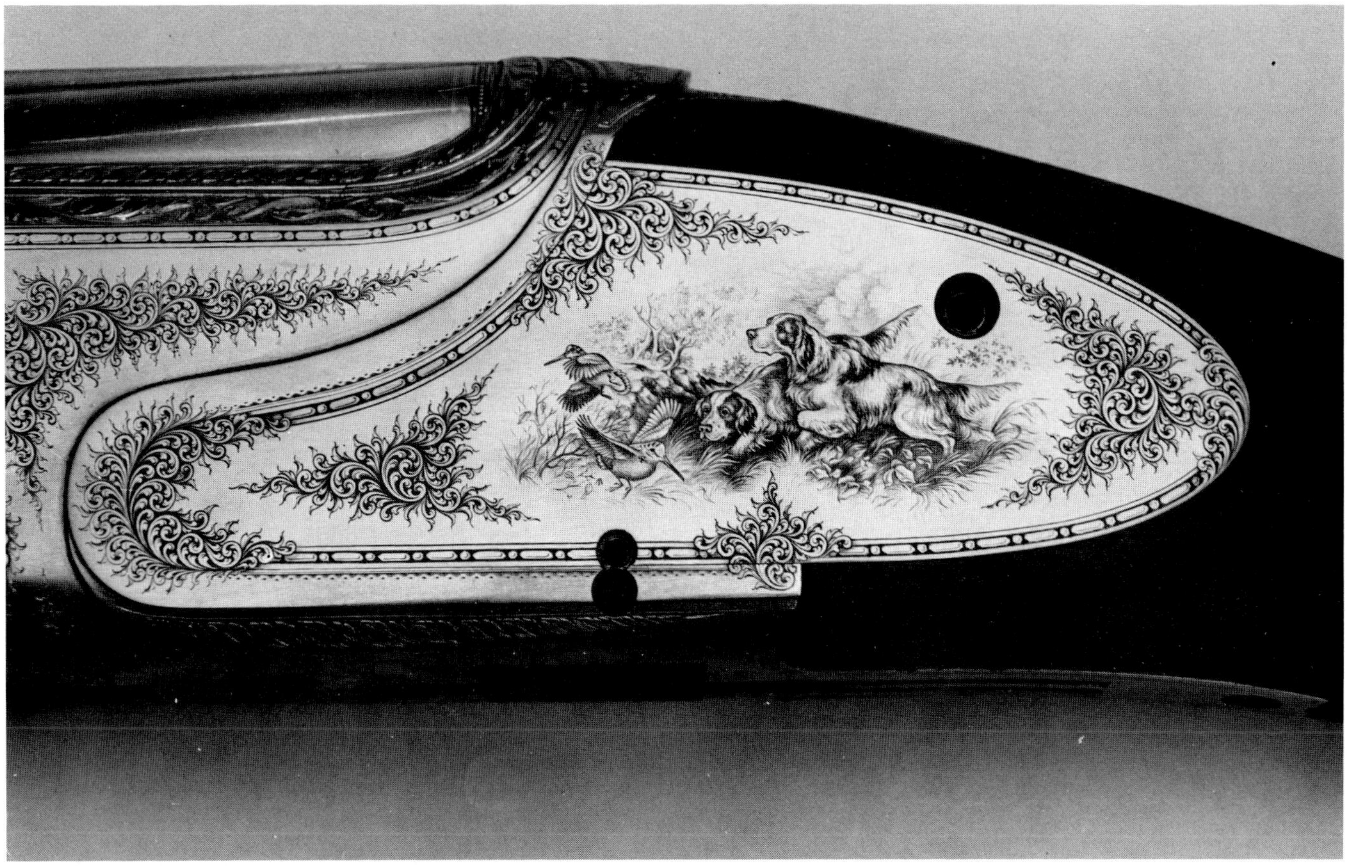

This photo shows Mr. Churchill's skill with English scrollwork and bank note style scenes. Many of the world's engravers use the term "Arabesque" interchangeably with the term "English scroll."

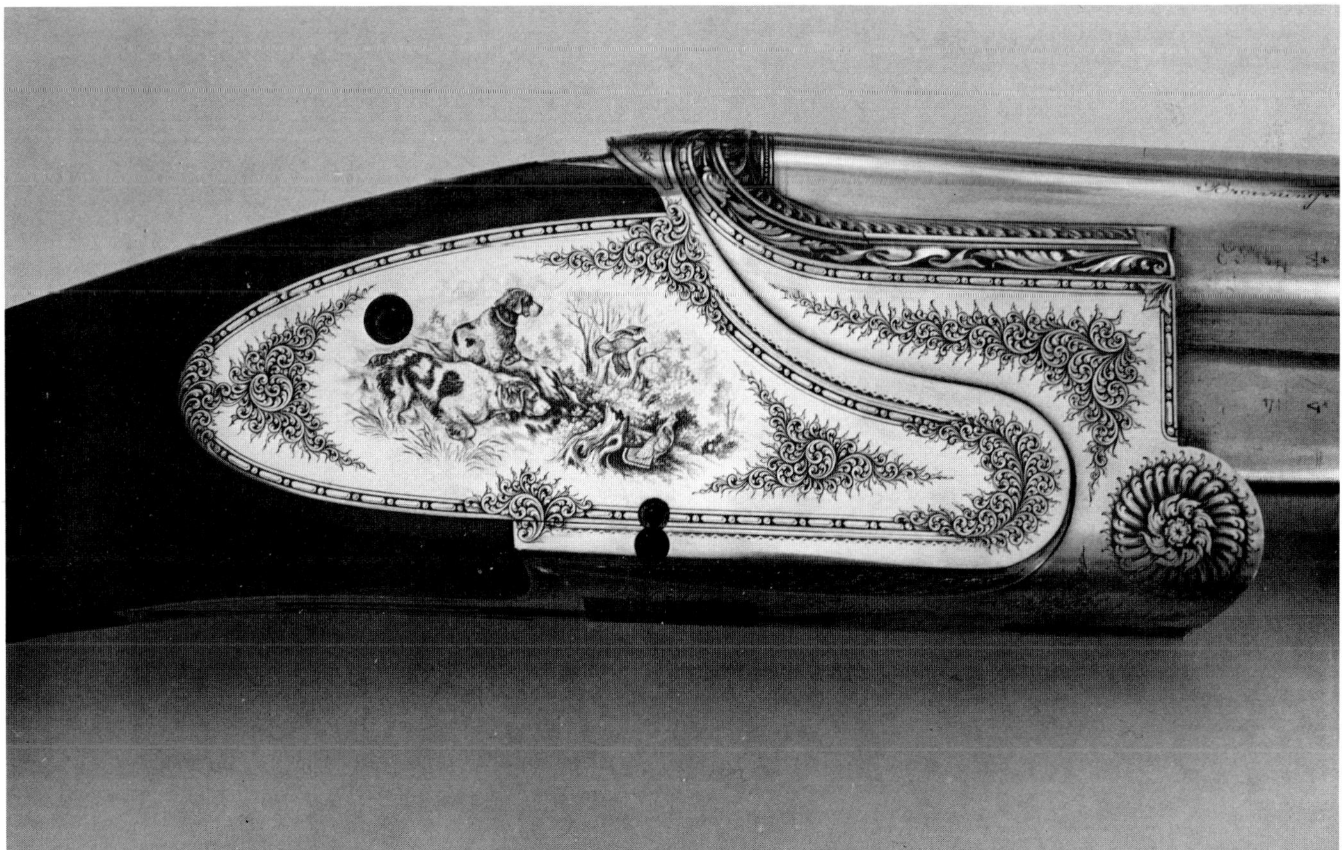

The sculptured steel design on the upper part of the receiver is displayed in this photo of a Browning Superposed shotgun. Also evident is the beautifully rosetted hinge.

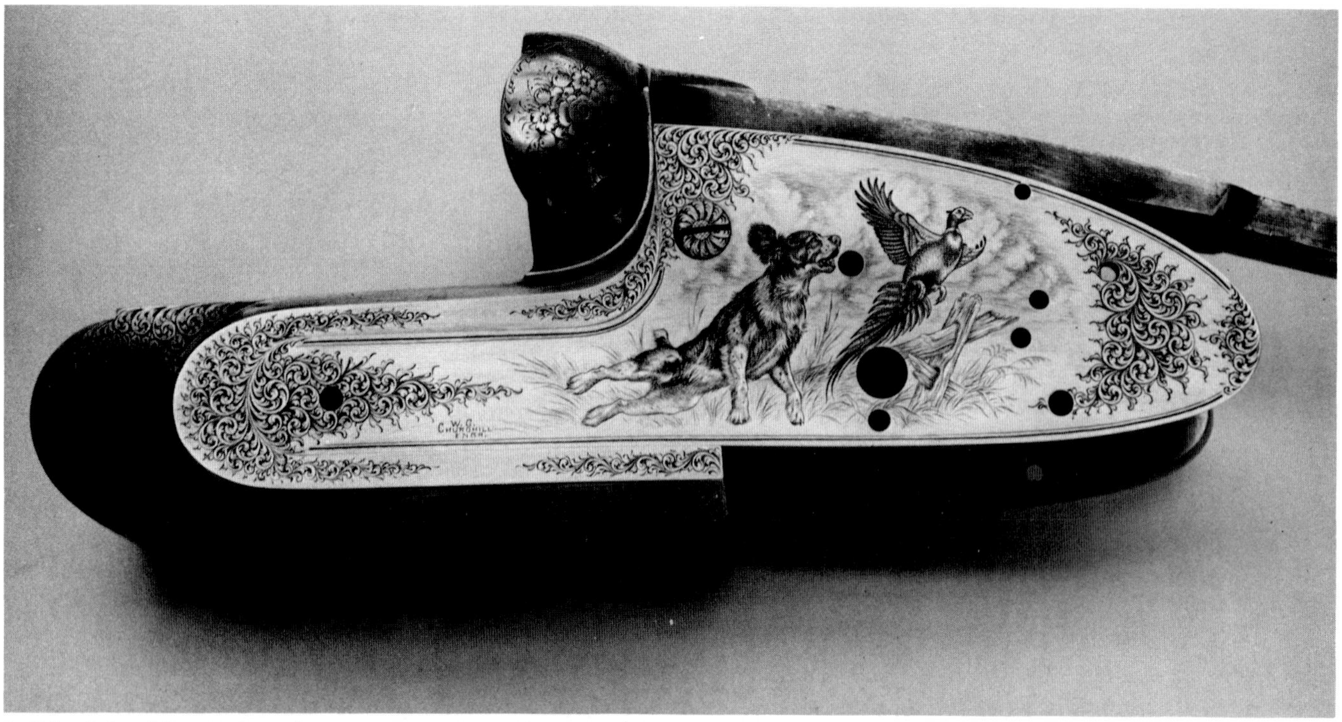

In this photo of the receiver of a side-by-side Purdy shotgun you can see Mr. Churchill's execution of English scroll and a bank note style scene contrasted with the original factory engraving on the fence, the sideclips, and the area of the hinge pin.

A variety of dogs and game birds on this trio of sideplates illustrates Mr. Churchill's talent at bank note engraving. Note the engraver's signature on each piece.

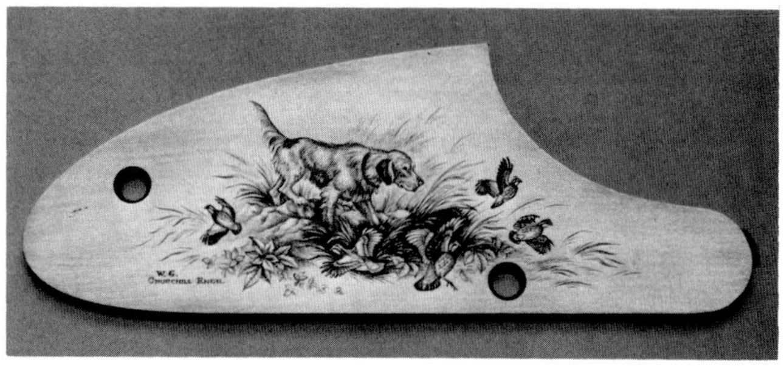

A native of Rochester, New York, Mr. DeMunck is a graduate of Monroe College, where he holds a degree in mechanical technology. Like many arms engravers his interest in firearms began when he learned to hunt and shoot as a youth, preceding his interest in engraving. Today this activity has broadened to include miniature arms. He has personally made and engraved two varieties of the Remington derringers in miniature.

Mr. DeMunck's professional career in the arms trade started in 1972, when he began work in the repair department of the Marlin Firearms Co. of North Haven, Connecticut. When he became bored with production line gun work, he accepted employment with the Pederson division of Mossburg Firearms, where he became an apprentice engraver. He has since returned to Rochester, New York with his wife and daughter, and is presently employed as a public safety dispatcher for Monroe County. He now engraves about fifteen hours per week, with hopes of more time at the bench in the future.

Mr. DeMunck feels that his style of engraving is slowly but continually changing for the better and hopes to develop an attractive style of scrollwork that can easily be identified as his own.

Mr. DeMunck currently accepts commissions from arms dealers as well as from private individuals. He is a life member of the National Rifle Association and the Genesee Conservation League of Rochester, New York.

# James R. DeMunck

*Rochester, New York*

**This Colt "Pocket Automatic" was originally purchased by Mr. DeMunck in very poor condition. To remove pitting and restore the metal surfaces, it was necessary to draw-file and hand-polish them. The lettering was recut and the grips were handmade.** (Collection of James R. DeMunck. Photo by Gene L. Davis)

**Detail of the Model 66 Smith & Wesson's right side shows a likeness of the owner's badge inlaid in relief gold and surrounded by Mr. DeMunck's own style of leafy scroll.** (Collection of Patrick M. Dellefave. Photo by Gene L. Davis)

**One of Mr. DeMunck's most recent works is this Model 34 Smith & Wesson revolver in .22 caliber. His bas-relief scroll with matted background has been accented by 24 karat gold borders and intertwining line motifs on the barrel and cylinder.** (Photo by Gene L. Davis)

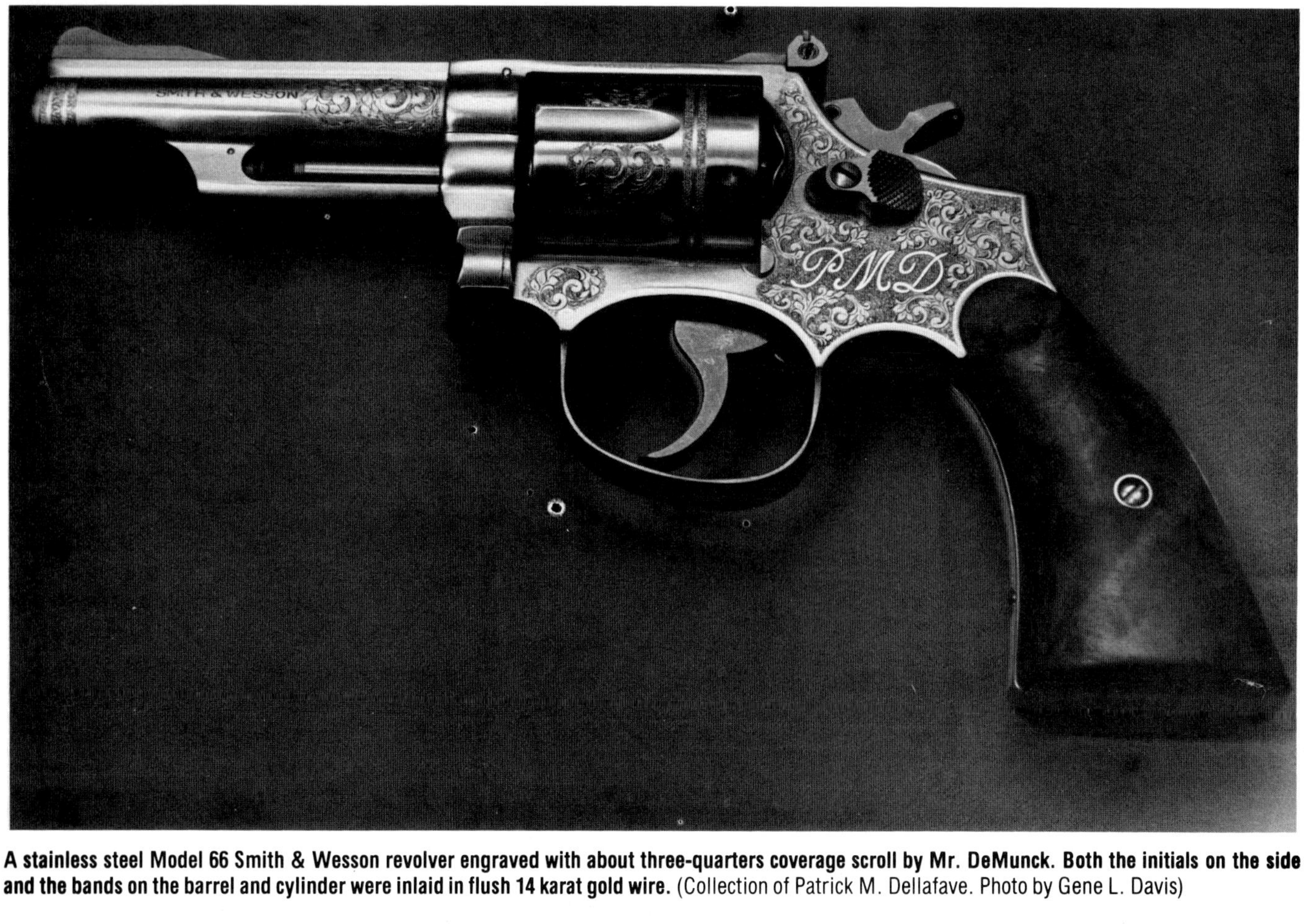


**A stainless steel Model 66 Smith & Wesson revolver engraved with about three-quarters coverage scroll by Mr. DeMunck. Both the initials on the side and the bands on the barrel and cylinder were inlaid in flush 14 karat gold wire.** (Collection of Patrick M. Dellafave. Photo by Gene L. Davis)

# Michael Dubber

*Evansville, Indiana*

A graduate of Indiana State University, Mr. Dubber is the holder of a bachelor's degree in fine art and a master's degree in technology. Until 1978 he was employed by the Evansville School Corporation.

Mr. Dubber became interested in the firearms field in 1965 and began doing part-time work with local gunsmiths, where he learned bluing, plating, and stock making. He also became very interested in the skill of engraving, but like many Americans he was disappointed by the lack of written material on the subject. Nevertheless he was able to begin doing some simple engraving by 1968. Since then Mr. Dubber has increased his skills to the point where he recently began engraving as a full-time occupation. He has also included some jewelry making and engraving in his activities, having studied under Ernest Tapscott, a jeweler.

In addition to private commissions he is presently engraving for Precision Shooting Equipment, Century Manufacturing, Inc., the maker of a unique .45-70 caliber revolver.

In the past three years Mr. Dubber has contributed three engraved Winchester shotguns to Ducks Unlimited for their fund raising projects. These guns were sold at Ducks Unlimited auctions in Indiana, Canada, and Louisiana, providing over $10,000 for the wildlife organization. He is active in his local chapter and is also a representative at national conventions.

Other interests include hunting, fishing, writing about the outdoors, and wildlife observation. His memberships also include the National Rifle Association, the National Muzzle-Loading Rifle Association, and the Masons.

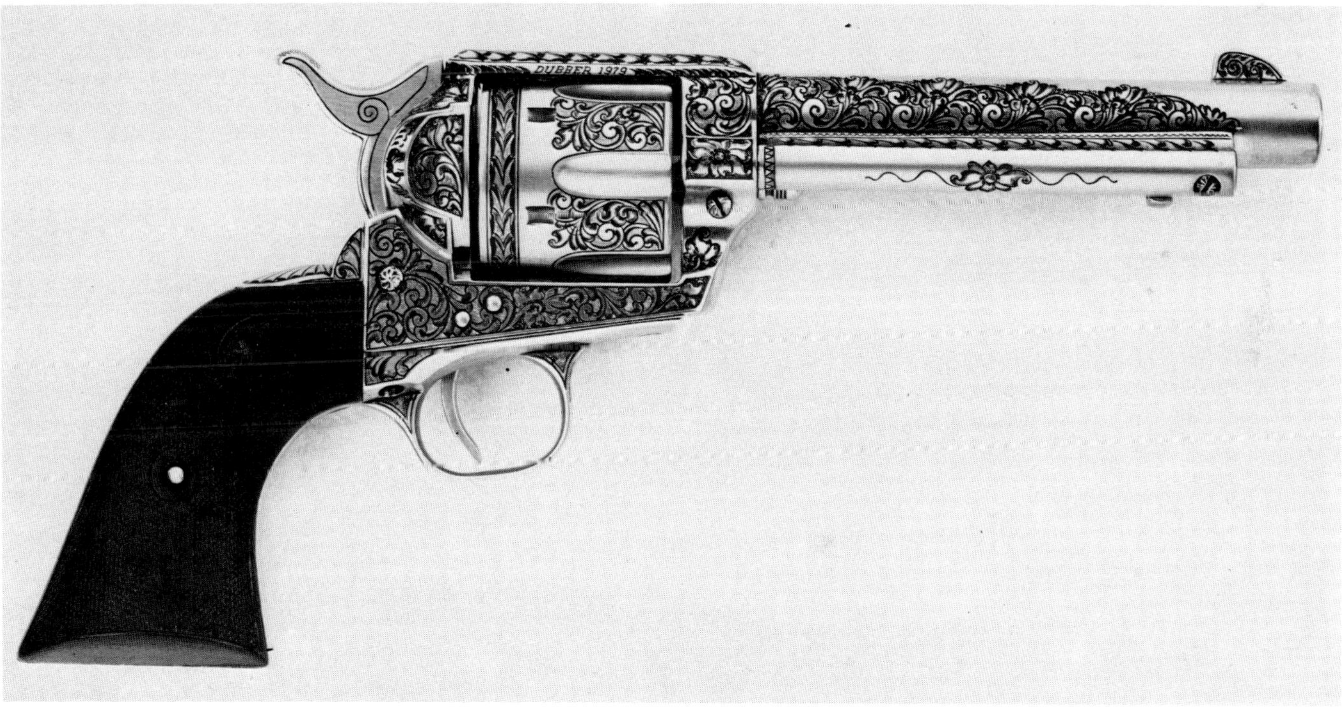

**This Colt Single Action Army revolver shows Mr. Dubber's style of American floral scroll used by him on most revolver commissions. Note the signature on the side of the top strap: DUBBER 1979.** (Collection of Robert Blemker)

This Model 12 Winchester has been engraved by Michael Dubber. The bas-relief scrollwork is surrounded by flush gold line that has been gone over with a wiggle tool. The water fowl are inlaid.

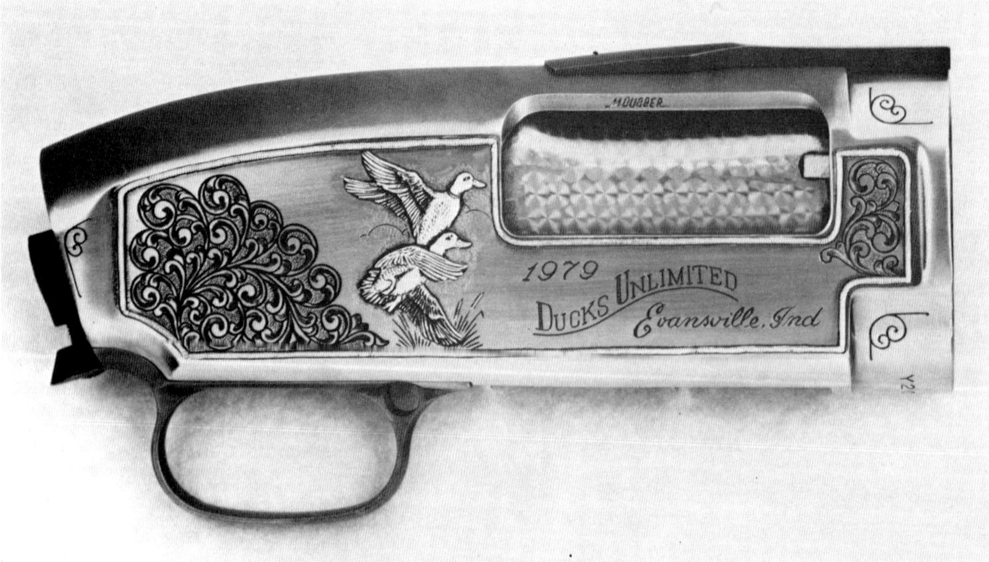


Photographed in the "white," the receiver of this Model 12 Winchester has been engraved for the Evansville, Indiana Chapter of Ducks Unlimited and will be sold at its annual auction. Note the signature on the ejection port: M. DUBBER.

Engraved in 1979, this Smith & Wesson has been treated by Mr. Dubber with American scroll, wiggle tool borders, and a bear's head motif.

# Robert Evans

*Oregon City, Oregon*

With degrees in history and education, Mr. Evans has been a high school history teacher for twenty years. His interest in firearms was a natural consequence of growing up with guns on an Oklahoma farm during the depression era combined with a strong interest in history, especially that of the American frontier.

He began collecting guns and building muzzle-loading rifles in 1958. His entry into the field of engraving began when he was unable to find anyone who could engrave the trim of his Kentucky rifles. When he located a silver engraver who was willing to give him basic instruction on the making and use of engraving tools, he took the first step on the long journey toward becoming a competent arms engraver.

In 1970 Mr. Evans gave up rifle building to concentrate on engraving. He later contracted to engrave a special series of knives for Gerber Blades. This project lasted two years. Since then he has concentrated his efforts on the embellishment of firearms.

Mr. Evans's experiences include five years of service with the United States Air Force as a flight engineer during both the Berlin air lift of 1948-1949 and the Korean War. He also worked and hunted in East Africa from 1963 till 1965.

Mr. Evans is one of several engravers who are members of the Masonic order. He is well-known to many collectors in the northwestern part of the United States because of his frequent attendance at the gun shows in the states of Oregon and Washington.

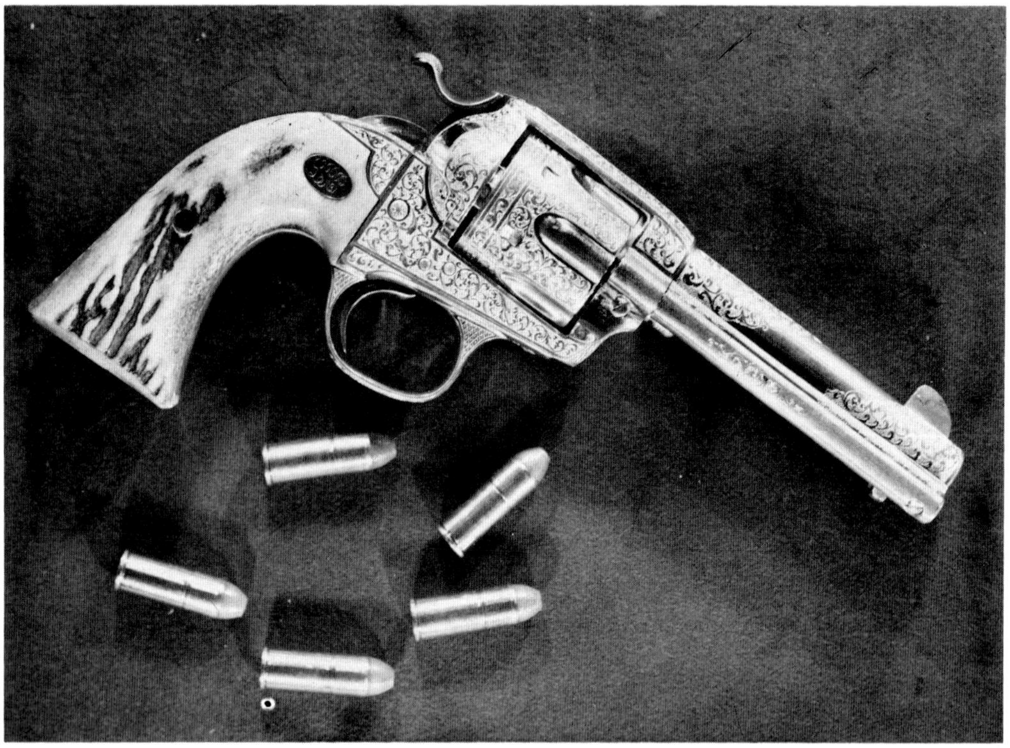

**This silver-plated Colt Bisley features stag grips with monogrammed inlays. The scroll designs are executed in bas-relief with circle-punched background.** (Collection of Les Pons)

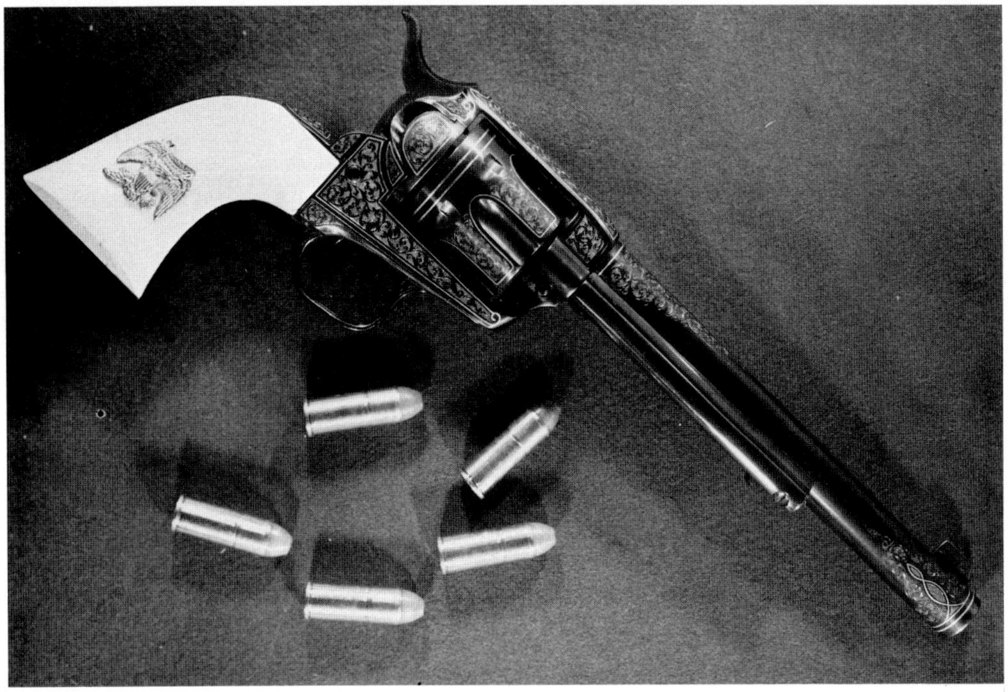

Engraved in relief scrollwork, this Colt six-shooter in .44-40 caliber also features ivory grips and 22 karat gold borders by Mr. Evans. Note the attractive intertwining line motif at the muzzle. (Collection of Daryl McFall)

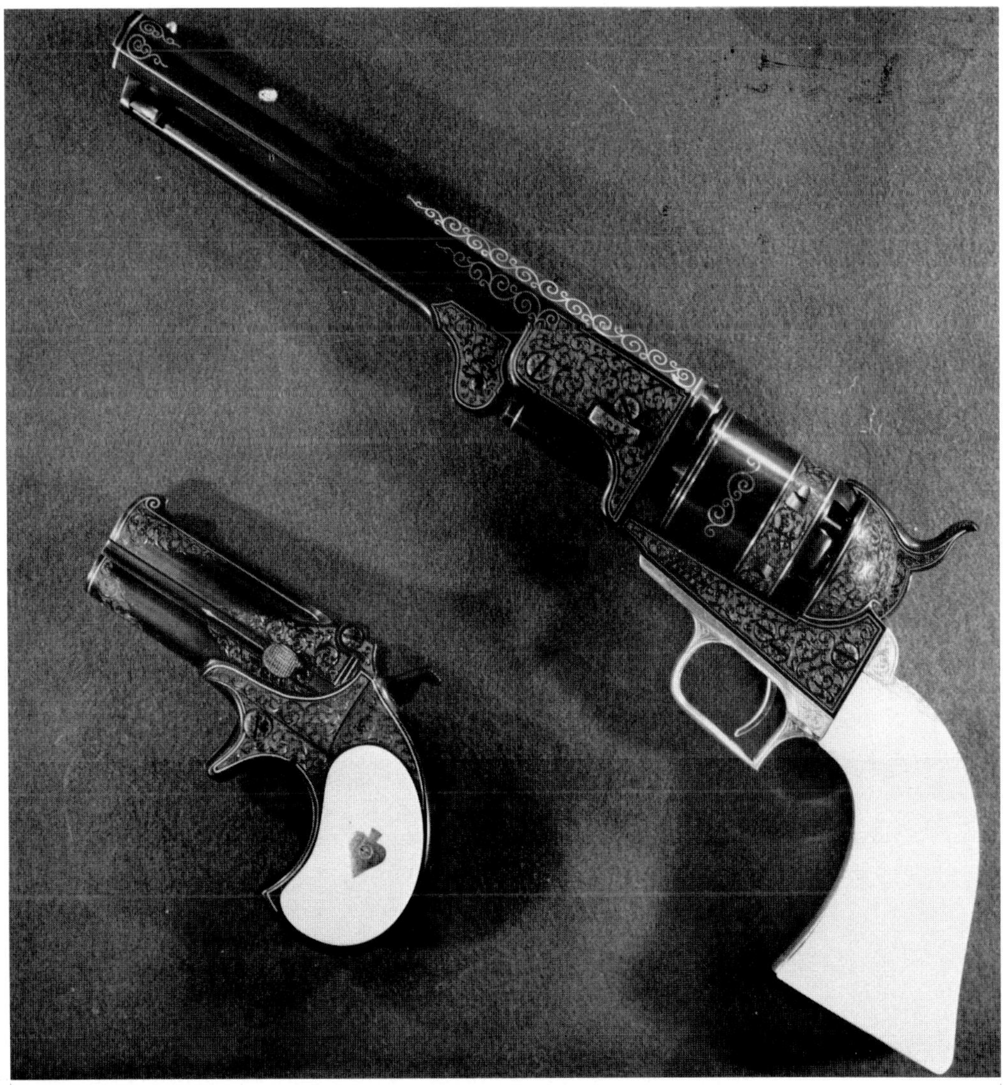

Extensive inlay of 22 karat gold wire is a prominent feature of these two guns, popular in the Old West. The Colt 1851 Navy is decorated with relief scrollwork, ivory grips, and lots of flush inlaid scroll on its octagon barrel. The Remington double derringer features a flush inlay of a spade design in its ivory grips, indicating the gun's popularity with western gamblers.

# Jaqueline Favre

*Las Vegas, Nevada*

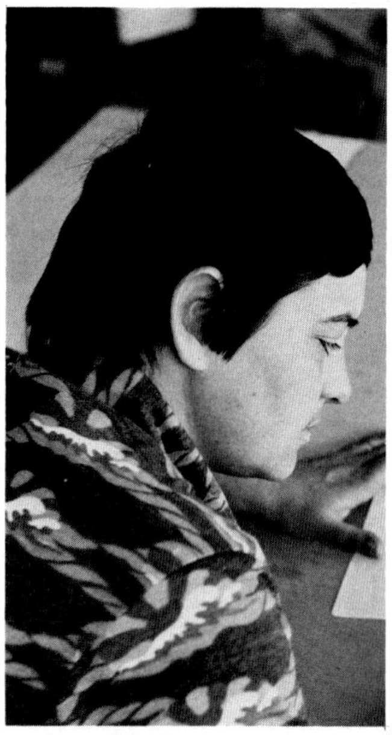

Born and educated in Switzerland, Ms. Favre came to the United States in July 1965.

While in Switzerland she had served a part-time apprenticeship at a stamp and seal shop in the town of Ostermundigen, where she had her earliest training in engraving. In America Ms. Favre studied at the School of Visual Art in New York City, while working in an office during the day. After a few years she moved to Las Vegas, Nevada, where she became the full-time protegee of William H. (Bill) Mains, whose work is also featured in this text. Ms. Favre counts Mr. and Mrs. Frank Sinatra among the patrons of her engravings.

The versatile Ms. Favre is an able translator of the German, French, Italian, and Baerndeutsch languages. She also enjoys target shooting.

A member of the National Rifle Association, the International Wildlife Federation, and the Defenders of Wildlife, Ms. Favre has been interested all her life in the beauty of nature. When not at her engraver's vise, she spends her time drawing, painting, and studying wildlife.

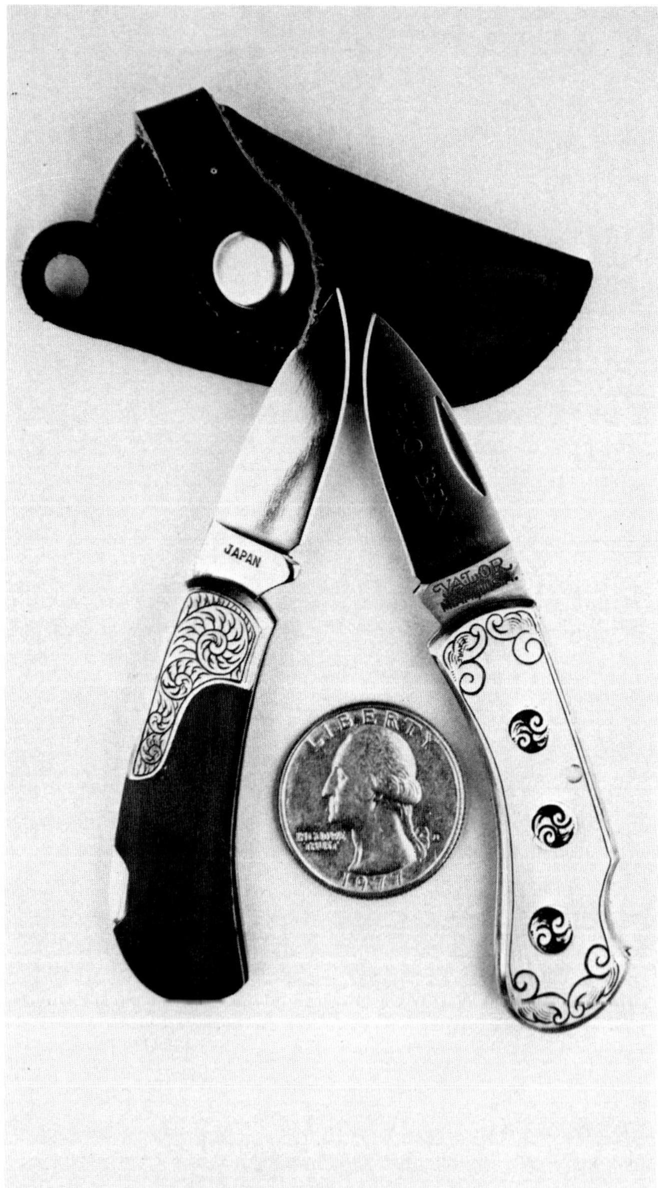


**These two miniature knives, shown with a quarter for size comparison, bear the work of Ms. Favre. The piece on the left is decorated with what most engravers call "English" scroll. Notice how each leaf inside the spiral touches another, with no background between.**

Ms. Favre has engraved the receiver of this Model 12 Remington rifle with deep relief decoration, which she refers to as "leaf scroll." The scene is line engraved. Note the signature, FAVRE, below the squirrel on the left.

Flush border inlays and light scroll decoration are features of this Astra vest-pocket pistol. The grips have also been carved with a scroll motif by Ms. Favre.

# Jeffrey W. Flannery

*Union, Kentucky*

Though born in Cleveland, Mr. Flannery grew up in the Cincinnati area of Ohio. He served his country in the United States Navy, aboard the U.S.S. *Holland*, with submarine duty in Rota, Spain.

After leaving the Navy he attended Ohio University, where he majored in business but also studied fine and industrial art. During school and after, Mr. Flannery received instruction in Kentucky rifle building from Cecil Brooks, a master at building these early style arms. He also worked with and learned from Jack Lewis, the Cincinnati area's most able and skillful gunsmith.

Later Mr. Flannery moved to Evansville, Indiana where he studied engraving for two years with Michael Dubber (featured earlier in this text). Mr. Dubber's influence is most apparent on Flannery's earlier works. But during the last three or four years he has tried to develop his own easily identifiable style.

In 1976 Mr. Flannery settled in Union, Kentucky, where he built a house and shop that overlook Gunpowder Creek Valley on the Ohio River.

**The side and frame of this "baby" Browning .25 caliber automatic are engraved with running vine American scrollwork and stippled background.** (Photo by Mike Van Oflen)

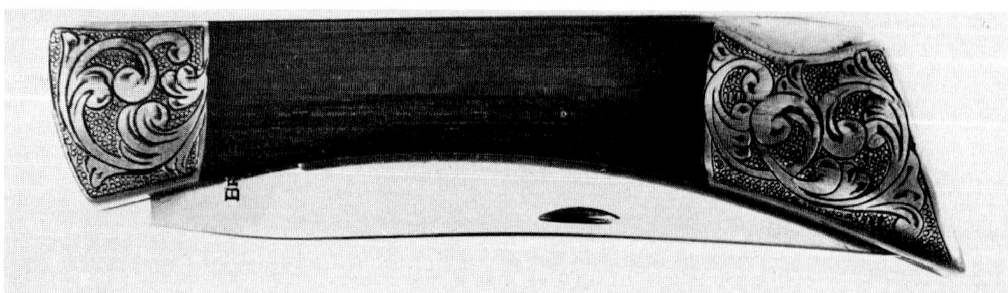

**Interlocking American scroll with a circle punched background was used by Mr. Flannery on this Browning folding knife.** (Photo by Mike Van Oflen)

These two views of a Model 12 Winchester skeet gun show Mr. Flannery's first attempt at inlaying figures on a firearm. Both figures are of silver. The surrounding ovals are inlaid gold that has been wiggle tooled.

# Guiseppi Forte

*Rawlins, Wyoming*

Relative newcomer to America's small cadre of arms engravers, Mr. Forte also works for the American Telephone and Telegraph Bell System in a technical capacity. In addition, he is licensed in Wyoming as a professional guide who specializes in antelope and deer hunts as well as wilderness fishing trips.

Mr. Forte credits John Rohner of Colorado with giving him valuable advice and instructions to improve his arms engraving skills. Rohner also introduced him to the use of the Gravermeister pneumatic-powered chisel, which he uses in conjunction with the ball vise (block). He states that he is still learning and improving with each task, and hopes to refine his engraving even more in the future.

An athletic man, Mr. Forte has played football and baseball and has participated in track meets. In the Navy he boxed for the Sixth Fleet in the light heavyweight class. He later began to shoot trap. Within two years he shot in the AA class with a 26½-yard handicap. He also plays golf in the low 80s.

**Similar in appearance to what Winchester calls a 12-1 pattern is this Model 12 shotgun. It is one of Mr. Forte's earliest efforts and features a gold inlaid goose on each side.** (Collection of Richard L. Valente. Photo by Dick Randall)

**This enlargement illustrates the multitude of cuts required for the graver to fashion a single line cut figure. The dot-like marks around the dog are made with a matting punch. The photograph shows a practice plate by Mr. Forte. The actual figure is ¾ inch high.** (Photo by Dick Randall)

# Burt, Francolini & White

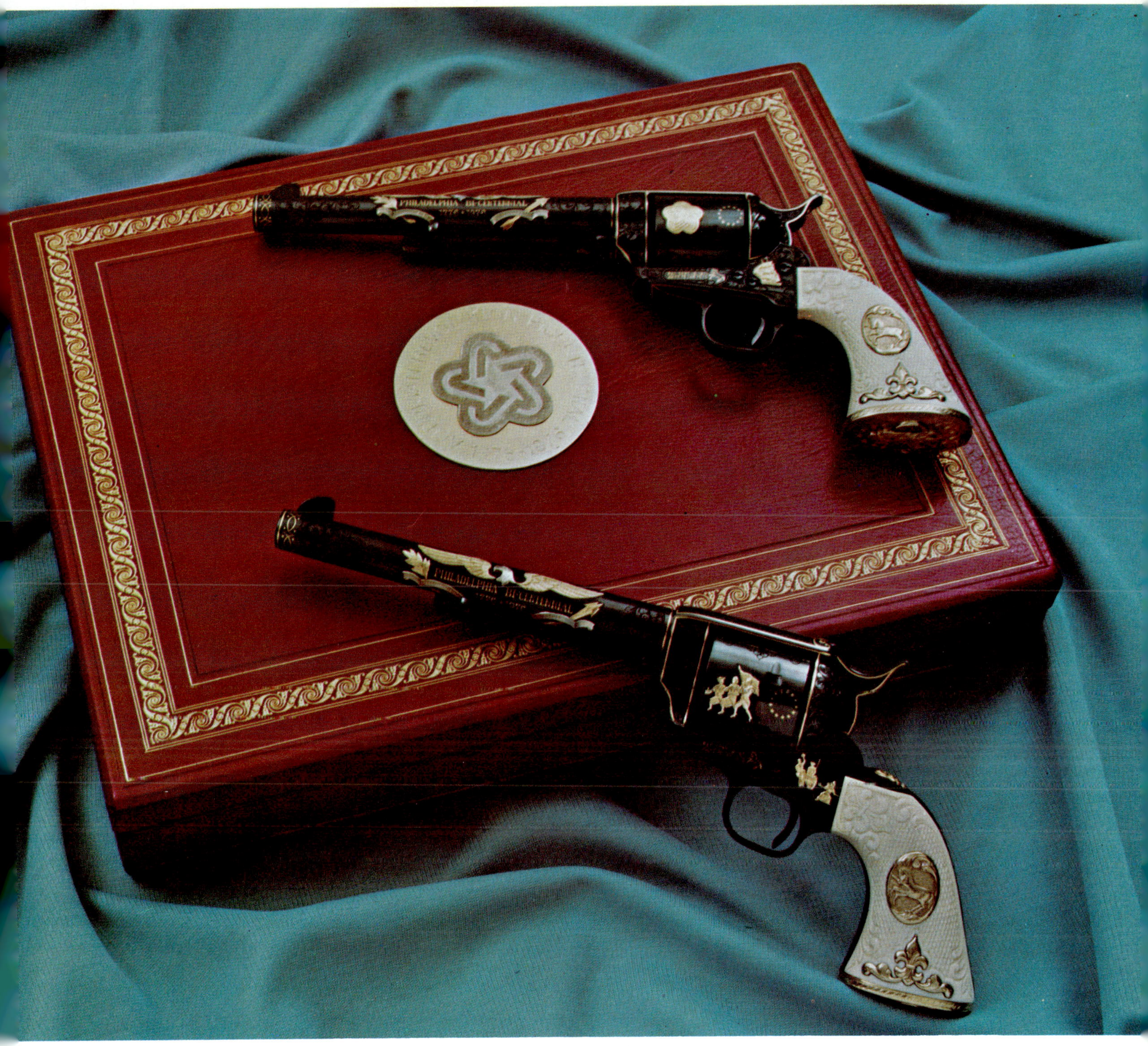

**Complete cased set of the Philadelphia Bicentennial Single Action Colts, engraved by Robert Burt, Leonard Francolini and A. A. White.** (Collection of Paul Frank, Jr. and Carl A. Braden)

Burt Sinclair engraved this Colt .32 caliber pocket auto pistol with full floral coverage.

# Jeffrey W. Flannery

**Engraved in a floral scroll style with matted background, this Browning Medallist target pistol has been given a gray finish that contrasts with the walnut stocks.** (Photo by Carl Bleile)

**The full coverage bas-relief scrollwork on this Browning 9mm "High Power" is typical of Mr. Forte's style. The piece is shown in "the white." It will later be given a frost finish and ivory grips.** (Collection of Philip D. Simpson. Photo by Dick Randall)

# Guiseppi Forte

**Mr. Forte's latest project is the engraved decoration on this Browning Model 78 single shot rifle. This rifle was commissioned by the One Shot Antelope Hunt held yearly at Lander, Wyoming. The scene on the left side depicts a Shoshone Indian hunting a buffalo with a lance, while Crowheart Butte is shown in the background.** (Photo by Dick Randall)

# Ken Hurst

**Mr. Hurst engraved this Colt revolver with a very tasteful amount of relief scroll coverage. The left side features gold inlays of a rampant horse and eagle's head; the right side is inlaid with a steer skull.** (Courtesy of Southwestern Shooter's Supply)

# Guiseppi Forte

**This .22 caliber Colt New Frontier revolver was the third firearm engraved by Mr. Forte. Done in bas-relief scroll, it features a large floral motif on the front of the frame and another floral pattern in a triangle on the recoil shield. The ivory grips are by John Suth of Indian Lake, New York. The gun is owned and used by the engraver's wife, Buttons.** (Photo by Dick Randall)

**A pair of custom-built presentation single action army revolvers. The grips feature sterling silver plaques which show Chuck Connors presenting a pair of single actions to Leonid Brezhnev at San Clemente. The deluxe embellishments, extra-long grips and gripstraps, and the stock inlays were executed by Mr. White, by order of Colt Firearms.**
(Photo by G. Allan Brown)

# Winston G. Churchill

Many of Mr. Churchill's talents are pictured here, including metalsmithing, stock making, toolmaking, engraving, inlaying, woodworking and cabinet making. This Model 53 Smith & Wesson revolver is fitted with two cylinders, one cylinder firing the .22 magnum cartridge, and the other firing .22 long rifle. The designs of the cylinder flats and frame are inlaid in sculpted relief gold. The barrel inlays, as well as the bordering of the frame and cylinder, are flush inlays. Note the Great Seal of the United States engraved in high relief on the top of the oil bottle. (Collection of Brown & Ives Trading Company)

**Liberty No. 1, a .45 caliber single action army commissioned by Colt as one of a limited number of deluxe single actions in the Great Moments in American History series. The gun features an inscription on the left side of the barrel, a gold relief buttcap of an American eagle, gold inlay on the barrel, an 18 karat gold front sight, gold scrolls and borders, gold barrel bands, and profuse scroll coverage. The backstrap is inlaid in gold script: Statue of Liberty Model, and the gun is further decorated with flush and relief gold inlays pertaining to the theme.** (Courtesy of Colt Firearms Division)

## John M. Kudlas

**This Ruger Blackhawk .357 magnum revolver has been profusely engraved and inlaid with thirteen figures in gold. Note the flush inlaid designs at the muzzle and rear of the cylinder. These designs are called Grecian frets or keys.** (Photo by Charles Twing)

Colt single action serial number 34982. This black powder specimen is elaborately engraved and gold-inlaid by Frank E. Hendricks, Jr. On the right side of the frame, a classic cavalry charge is dramatically gold encrusted. The gold inlay on the left recoil shield is of General George Crook and the cylinder flats are inlaid with cavalry accouterments. (Collection of Buddy Hackett)

# Acevedo & Haas, Sr.

Colt single action serial number 112222 honors the lawmen of the west. Representing the work of Mr. Acevedo of Thomas Haas's Guns Unlimited, this gun features a gold-inlaid bust of Wild Bill Hickok carved in relief on the recoil shield. The gun also displays engraved and inlaid scenes of Wild Bill's famous gunfights, as well as scenes of his days as a stagecoach driver, a Union Army Civil War scout and a wild west showman. The ejector gate housing is inlaid with a famous poster of Wild Bill's showman days. (Collection of Buddy Hackett)

**The Colt Bisley model pictured here has been engraved by Jim Kelso in what he calls "Nimschke style" scrollwork. He has incorporated a wheat chaff pattern at the rear of the cylinder and a star motif on the ejector boss of the frame. The color case hardening and bluing is by George McKay.** (Collection of Christopher Klitgaard. Photo by Mustafa Bilal)

# Alvin A. White

Decorative devices on this Colt single action represent the personal enthusiasms of the owner. They include an architectural device on the cylinder (the owner's front door), a patriotic American eagle on the recoil shield, an African lion on the barrel, an acanthus leaf and the Campfire Club of America logo in relief gold on the backstrap, the spine of a book in relief gold on the buttstrap, the Ferrari and Mercedes emblems in engraved flush gold on the butt of the grips, an inlaid serial number on the frame, the owner's name inlaid in flush gold script on the backstrap, as well as a profusion of other scroll-engraved and gold-inlaid devices. (Photo by Sid Latham)

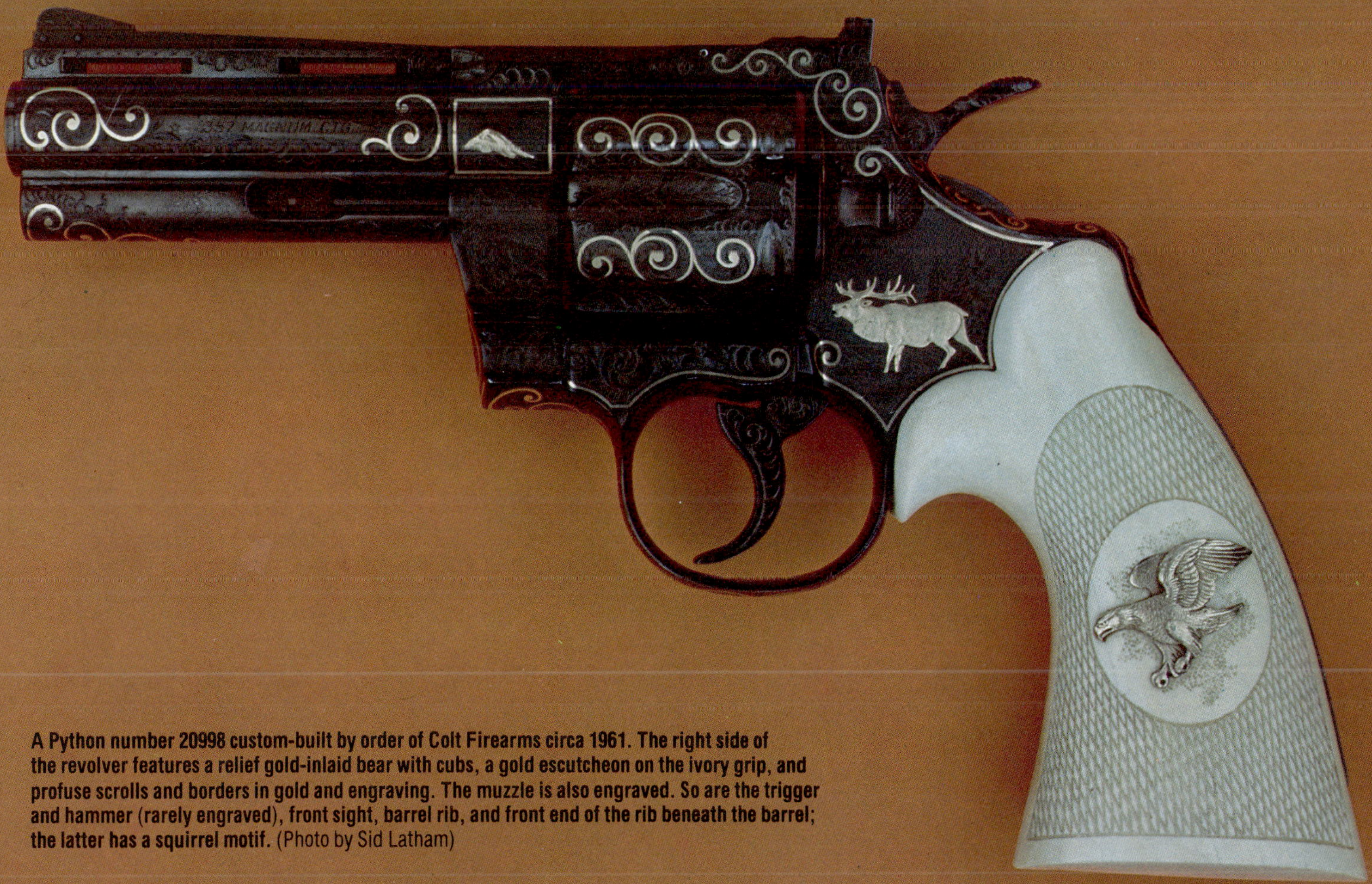

A Python number 20998 custom-built by order of Colt Firearms circa 1961. The right side of the revolver features a relief gold-inlaid bear with cubs, a gold escutcheon on the ivory grip, and profuse scrolls and borders in gold and engraving. The muzzle is also engraved. So are the trigger and hammer (rarely engraved), front sight, barrel rib, and front end of the rib beneath the barrel; the latter has a squirrel motif. (Photo by Sid Latham)

# Leonard Francolini

Mr. Francolini's talent at flush and raised inlay is illustrated here on this Colt Navy model revolver. Though the backstrap and hammer are yet to be decorated, the finished parts of the gun reflect a style of work that Mr. Francolini once applied to Colt's finest exhibition pieces. (Collection of Leonard Francolini. Photo by Jeff Yardis)

# Leonard Francolini

*West Granby, Connecticut*

Among arms collectors is a stereotyped image of the master engraver: an elderly gentleman whose German accent is as thick as his eyeglasses. Though many master engravers have in fact fit this image, Leonard Francolini is one of a number of fine, young, and native-born American engravers.

Born in Connecticut, Mr. Francolini began his engraving career after college. As a die cutter with Preston Engravers of Windsor, Connecticut, he practiced the basic disciplines of hand engraving and learned the properties of steel.

After three years as an industrial engraver, Mr. Francolini discovered the work of nineteenth century master engraver L. D. Nimschke at the Wadsworth Atheneum. From then on, he strove to acquire the additional skills necessary for arms engraving.

Assisted over the phone by Colt factory engraver Bob Burt, Mr. Francolini began as a self-taught gun engraver. Hired first as an apprentice in 1971, Francolini made a career at Colt, where he worked on many of the company's most lavish projects, including the famous Bicentennial Dragoon which originally sold at auction for $55,000.

In August of 1979, after eight years with Colt, Francolini began working as a free-lance artist.

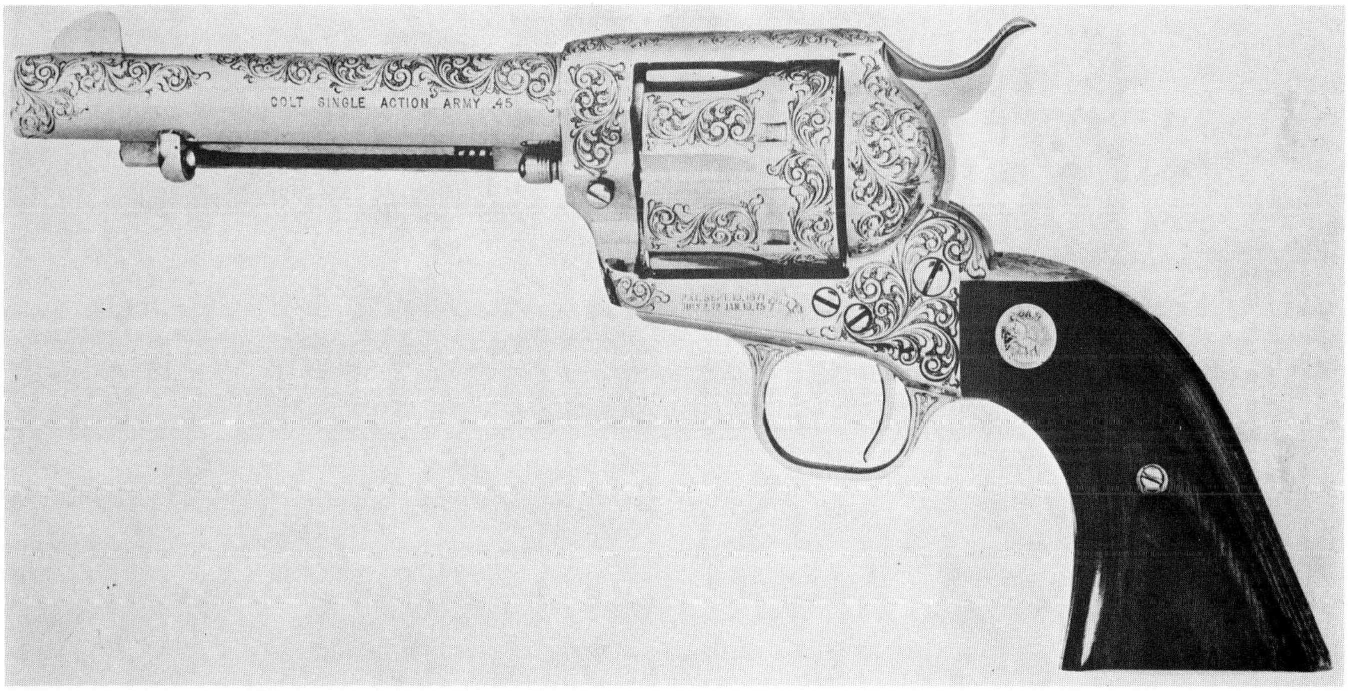


**The Colt Single Action Army shown here illustrates the engraving done early in Mr. Francolini's career. This piece was used in Colt's 1974 catalog to illustrate "C" coverage engraving.**

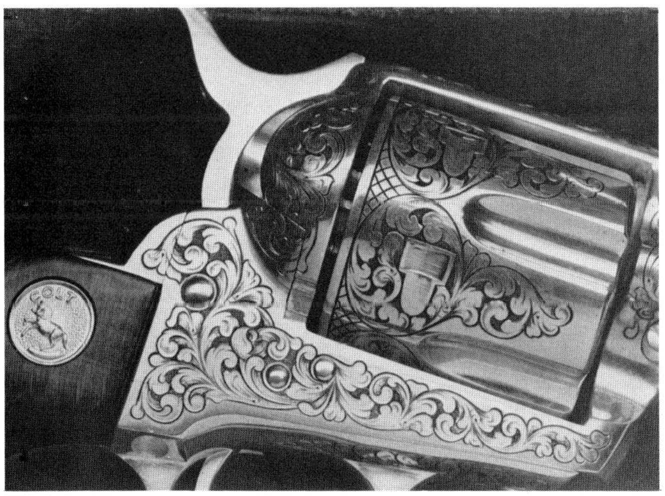

**A close-up of the frame and cylinder of this Colt show Mr. Francolini's relief scrollwork executed on all but the back of the cylinder without the restrictions of border lines.**

**The left side of this pre-World War II Colt vest pocket pistol shows Mr. Francolini's work in raised and sculptured gold inlay.** (Photo by G. Allan Brown)

**Leonard Francolini has incorporated bas-relief scrollwork with raised and flush gold inlays on this Navy model Colt percussion revolver.**

**Enlarged view of a Colt Navy cylinder showing flush inlaid figures bank note style.**

The high relief scrollwork shown on this Colt Gold Cut target pistol is typical of the style used by Leonard Francolini on most of his recent works.

## Heinrich H. (Henry) Frank

*Whitefish, Montana*

Henry Frank came to America in 1951 from his native West Germany. He was born in 1931 in a small town near Hamburg, where he attended school. He also graduated from a four-year technical school, where he majored in engineering, metallurgy, and design.

Before going into his own business, Mr. Frank worked in various metal-oriented fields such as electric welding, tool and die making, and tool designing. For thirteen years before his "retirement" he worked as a design engineer, and holds ten United States patents in the mechanical field. During this period he kept a workshop in his home where he would occasionally engrave or repair a gun. The arts are one of his first loves. Mr. Frank has done oil painting and wood carving and has made stained glass lamps. He has even made ships in bottles.

In 1961 Mr. Frank designed, made, and engraved a folding knife when he became dissatisfied with the knives he saw for sale. This first knife was seen by many people and it created a demand for his edged creations.

Since his retirement from the machine tool industry he has moved to Montana, where he has become one of the foremost makers of handmade folding knives in the world today. At present, he does not accept commissions to make knives because he has a twelve-year backlog of orders; nor does he usually accept engraving work as he prefers to work now only on his own knives.

Mr. Frank says that his income is now much less than when he was an engineer, but thanks to his beautiful surroundings in Montana and to working at the things he likes he enjoys life more than ever.

**A Walther PPK, appropriately engraved with high modeled relief oak leaves and acorns by Mr. Frank. The gun is gray finished, with ivory grips also fashioned by the engraver. The opposite side of this gun is pictured in the 1977 edition of the** *Gun Digest*. (Photo by H. H. Frank)

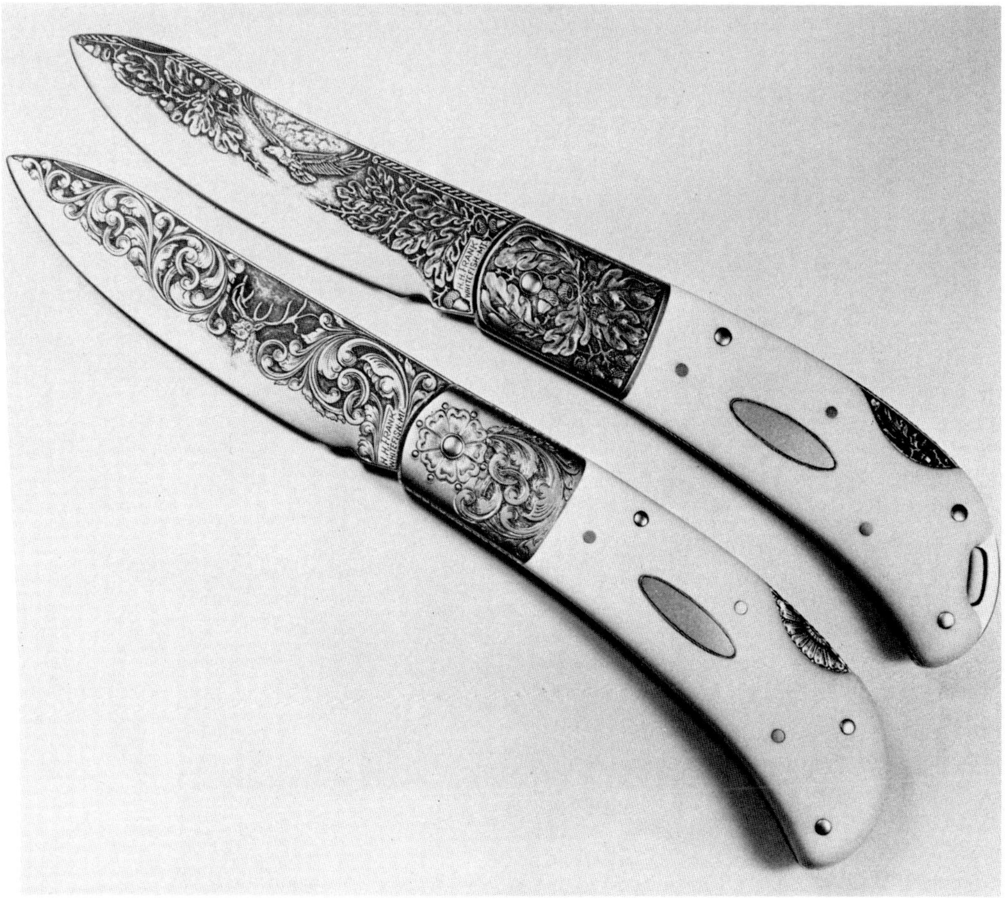

Two of the handmade folding knives that Henry Frank is well-known for making and engraving. Note the signature on the tang of each blade: H. H. FRANK / WHITEFISH-MT-.

**This triggerguard and floor plate assembly has been engraved by Mr. Frank with the likeness of a bighorn ram and surrounded with scroll and decorative motifs in high relief. Note the relief monogram FC on the trigger bow.** (Photo by H. H. Frank)

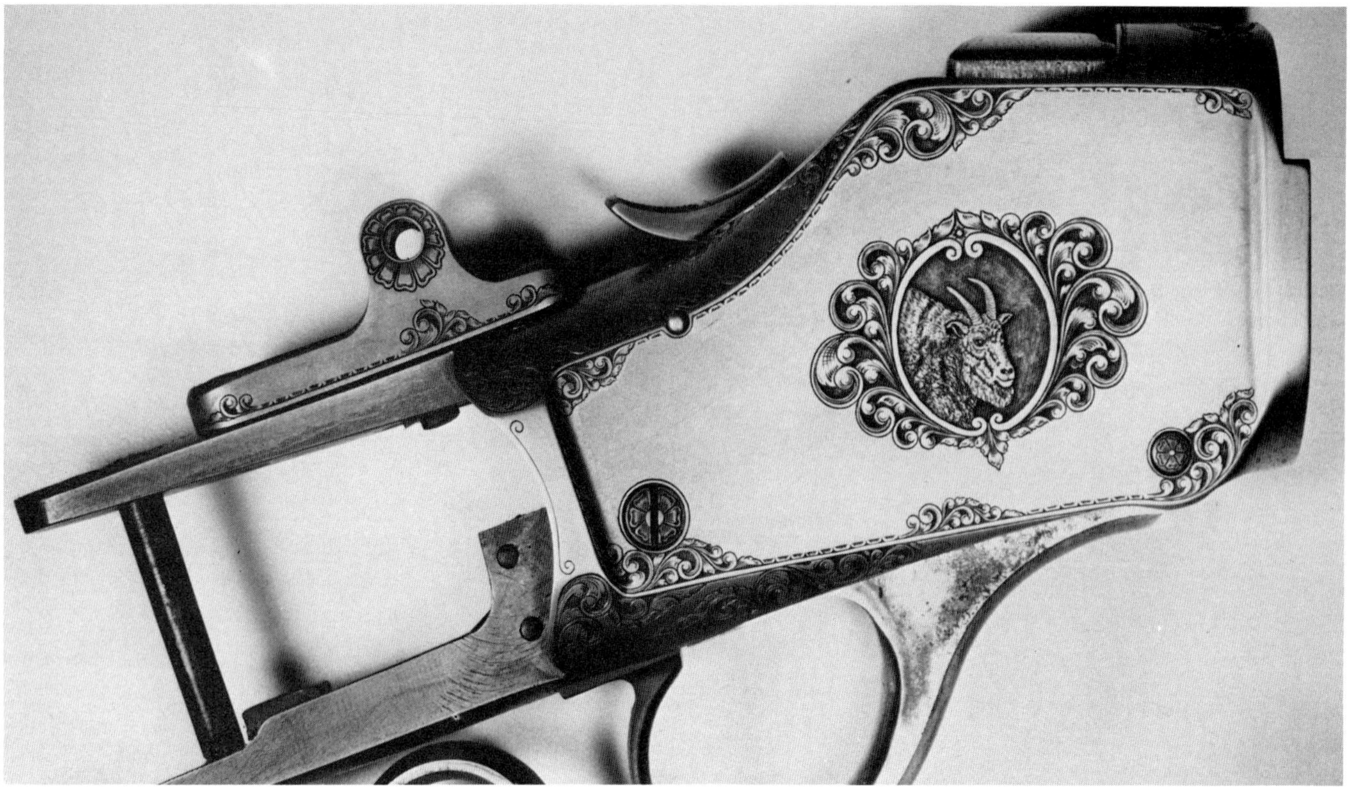

**The receiver and sight base of this Winchester "High Wall" single shot rifle has been tastefully engraved with high relief scroll and game animals. Note the flower petal decoration on the screw heads and the engraver's signature on the left side of the action: 19 ENGR. H. H. FRANK 79.** (Photo by H. H. Frank)

# Daniel Goodwin

*Kalispell, Montana*

One of several American engravers once employed at the Colt factory, Dan Goodwin began his career in 1973 when he studied jewelry engraving at the Bowman Technical School. After graduation he was hired as an apprentice engraver at the Colt factory, where he rapidly progressed to master status. During his tenure at Colt, he worked on four of the company's lavish N.S.G.A. show guns, as well as on many special commemorative and custom pieces.

Early in 1979 Mr. Goodwin left Colt to work as a free-lance artist. He now has a shop in Kalispell, Montana, where several other custom gun makers also reside. Among his commissions as a free-lance engraver was serial number 3,000,000 Marlin lever action rifle which was pictured on the cover of the August 1979 issue of the magazine, *American Rifleman*.

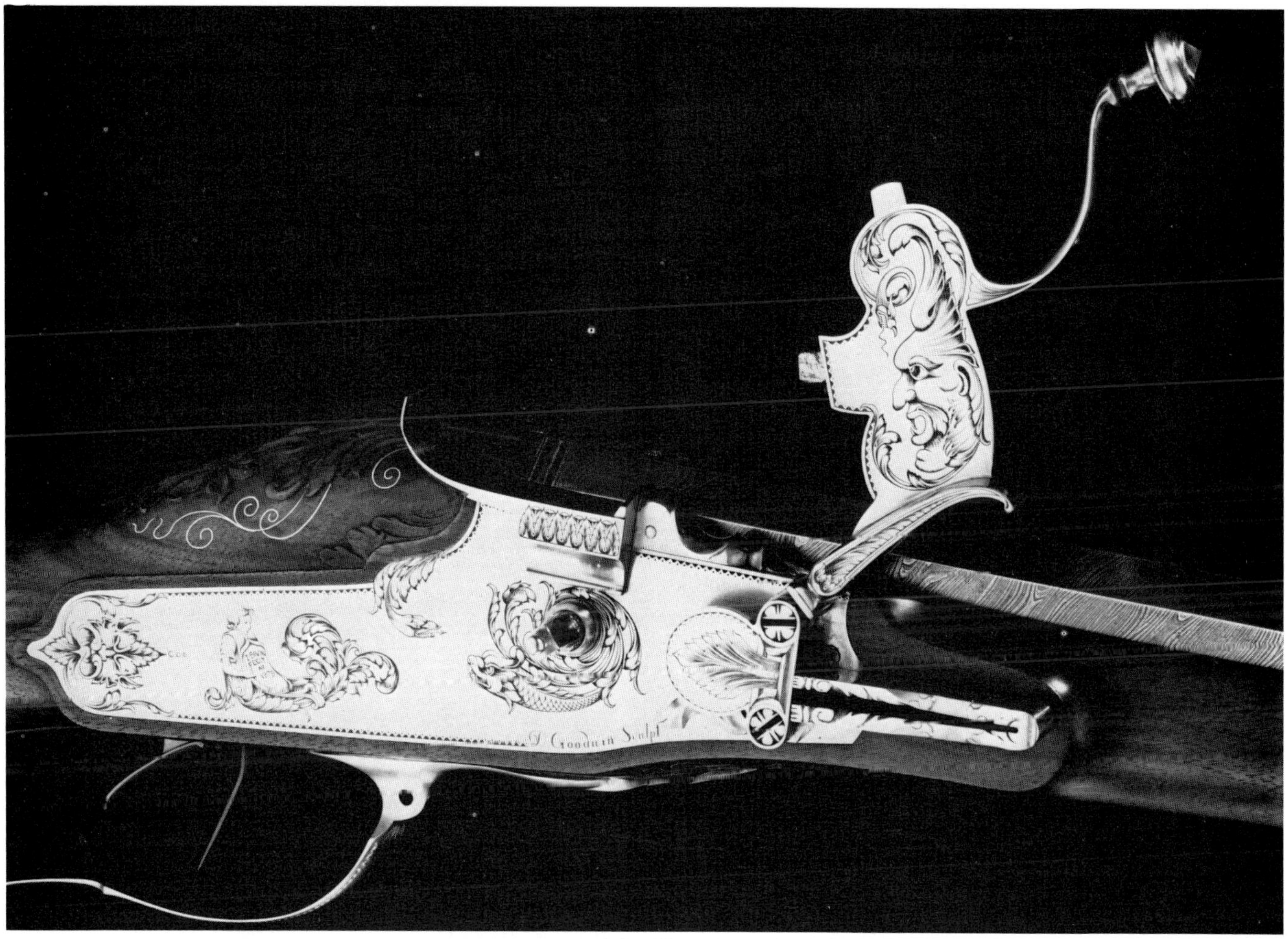

**Pictured here is a wheellock rifle completed in 1979 by gun maker John Bivens. This piece was designed to conform to the Nuremberg style of about 1690. Mr. Goodwin has designed and executed baroque decoration along the lines of the Paris school of the late seventeenth century. This style is known to Austrian and German engravers as Barockgrotesken, Flachstich. Note the signature, D. Goodwin Sculpt. Sculptor is an early designation for engraver.** (Photo by H. Armstrong Roberts III)

**Engraved in a contemporary vein is this brass floor plate made by Ted Blackburn. Cut in flat relief, Mr. Goodwin has executed a pronghorn antelope surrounded by a horse-shoe-shaped cartouche.** (Photo by Dan Goodwin)

**Mr. Goodwin has engraved a baroque motif on this sideplate made by Monte Mandarino for a German long rifle in the style of the early eighteenth century.** (Photo by H. Armstrong Roberts III)

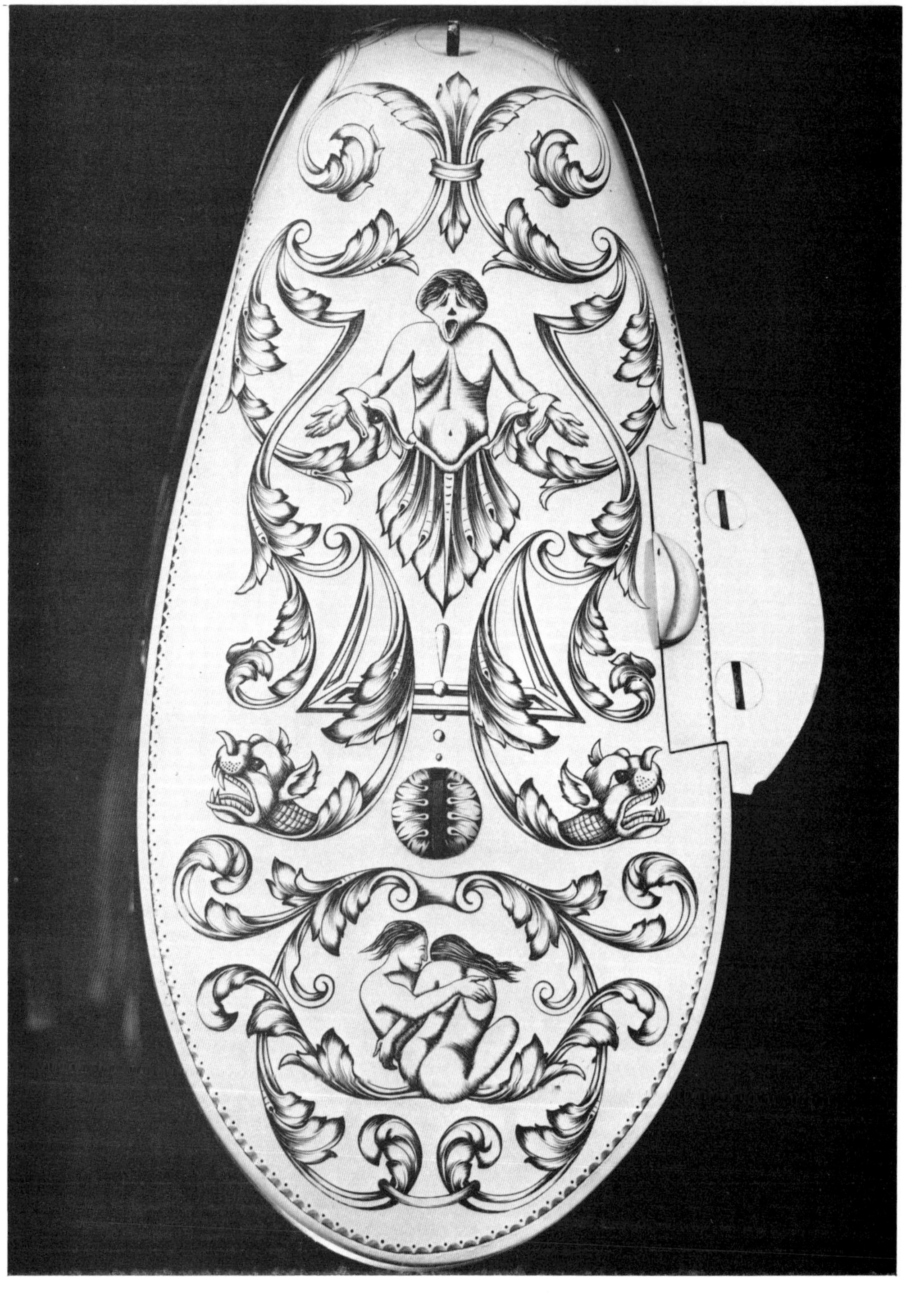

**Detail of the buttplate of a Bivens wheellock rifle. To achieve a three dimensional effect, the baroque engraving seen here requires great finesse in shading.** (Photo by H. Armstrong Roberts III)

# Howard V. Grant

*Lac du Flambeau, Wisconsin*

Born in Elgin, Illinois, Howard Grant has been involved in art and engraving for most of his sixty-three years. He cultivated his interest in art at the Academy of Fine Arts in Chicago, where he studied from 1937 to 1939. Then he spent the next five years as an apprentice engraver with the Elgin National Watch Co.

For the next twenty-five years Mr. Grant worked in various phases of the watchmaking trade, at different locations throughout the United States. During those days he engraved special presentation watches for many notables, among whom were Winston L. Churchill, Franklin D. Roosevelt, and Joseph Stalin. The Royal Family of England has also received his work.

When Elgin moved to New York in 1970, Mr. Grant moved his family to a new home in Wisconsin. There, because of his friendship with gunsmith Norbert (Nubs) Fashingbauer, he tried his hand at engraving firearms. Although the transition from engraving soft precious metals to the hardened steel of firearms was difficult, he found that he enjoyed the change.

Since 1970 Mr. Grant has engraved many firearms and related items. He has worked in several styles of decoration as well as with inlays. His specialty is engraving detailed wildlife scenes, many times in high relief. He accepts work from private individuals only and, like most American engravers, has no employees or apprentices.

**Engraved in 1974, this 6- by 6-inch sterling plate was later mounted on red velvet and framed. Note that the bas-relief scrollwork has been backgrounded with line cuts by a single point graver.**

**By eliminating the traditional scrollwork borders, Mr. Grant has achieved a sort of wraparound effect on the underside of this Model 21 Winchester side-by-side shotgun.** (Collection of George Martens)

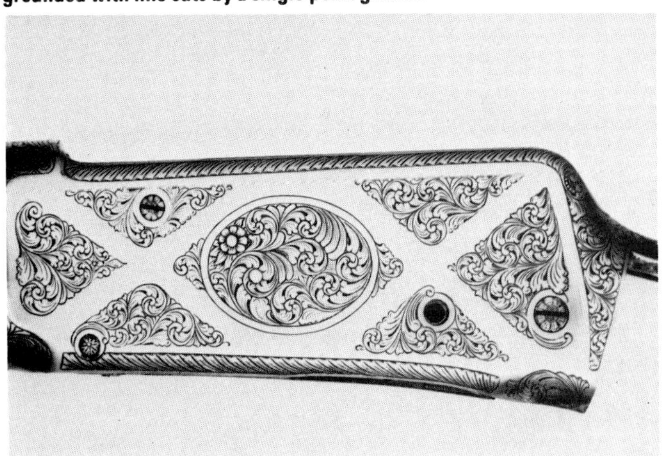

**The idea of using scrolls cut to form islands that result in a design with the unengraved surface is a popular motif on English guns. Note the "Bat Wing" motif used as borders along the top and bottom edges of the frame.** (Collection of John Todd)

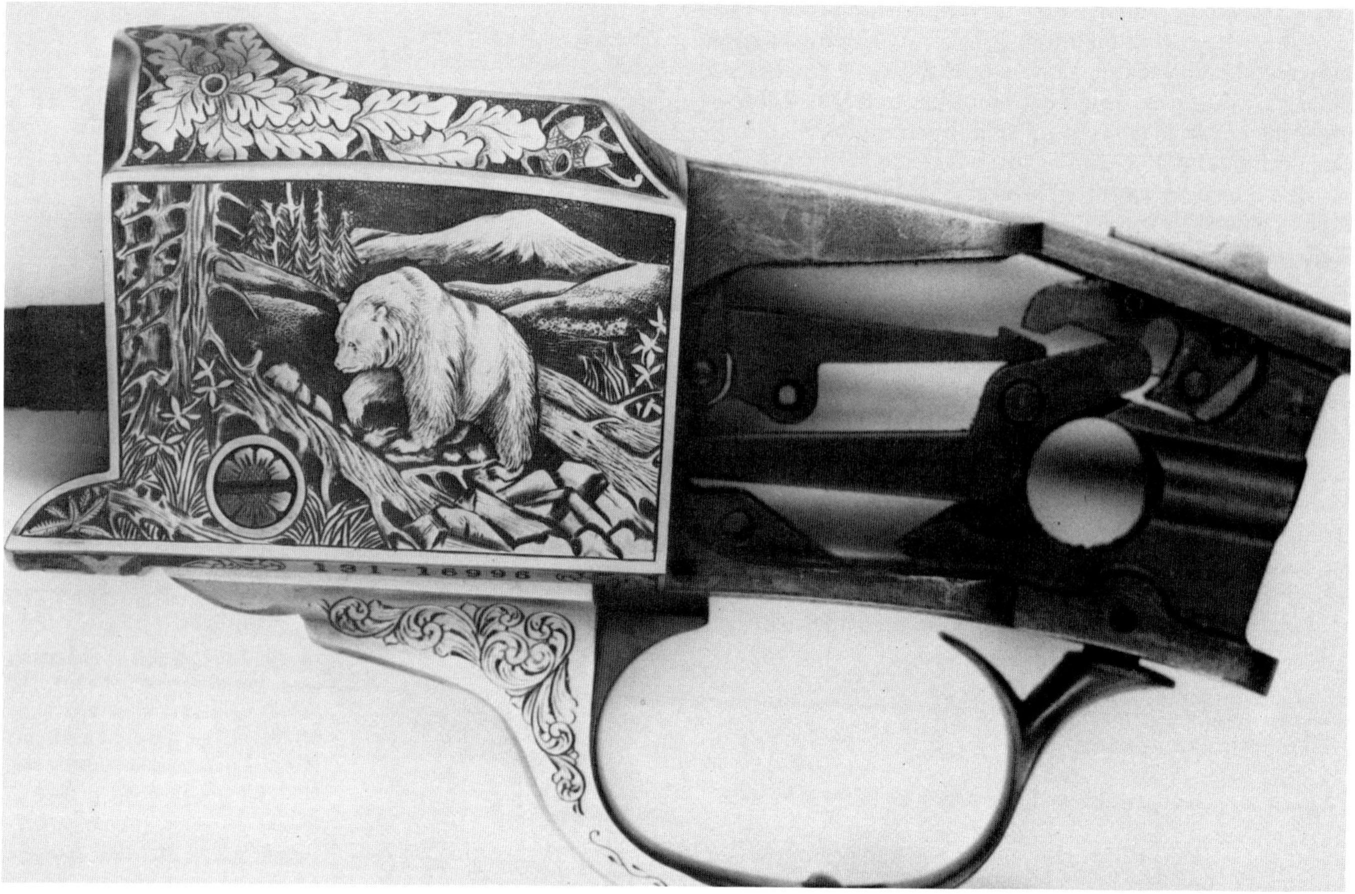

Mr. Grant has achieved a unique effect from filling the entire side surface of this Ruger No. 1 (in .458 caliber) with game scenes having no decorative borders. A further innovation is the use of oak leaves on the top of the receiver and large English scroll on the lever. (Collection of Robert Martin)

**The high relief game scenes on the receiver of this .375 caliber Ruger No. 1 rifle were suggested by two oil paintings. The left side (above) was inspired by "The Contested Game," circa 1879, by William de la Montagne Cary. The right side (below) was inspired by "Crossed Trails," circa 1879, a painting by John Clymer.** (Collection of John Selleck)

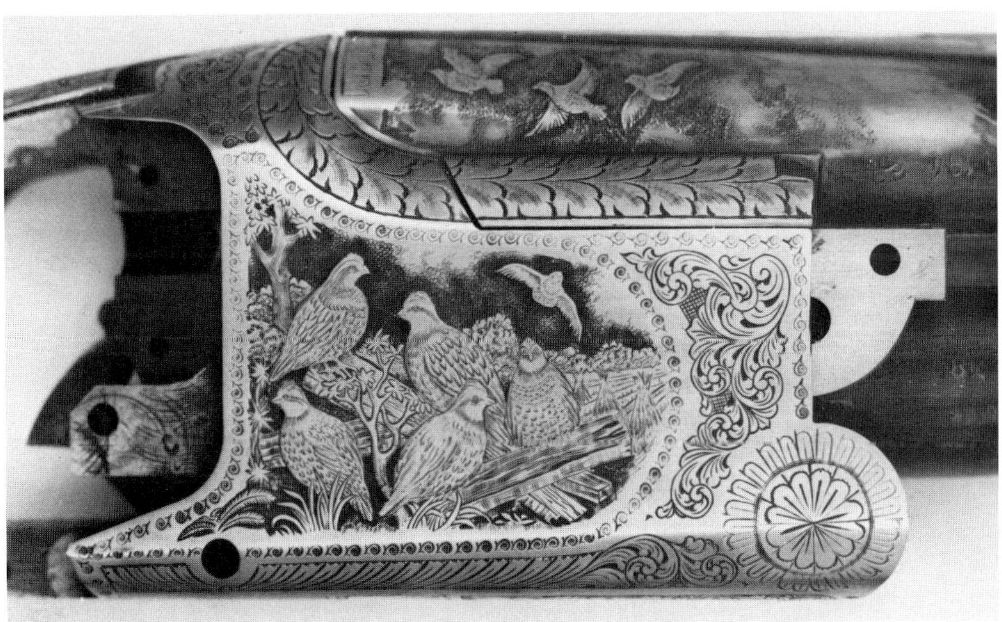

**This Browning Superposed was engraved with relief game scenes suggested by a print by wildlife artist Owen Gromme.** (Collection of Harlow Scherf)

# Fred Alan Harrington

*Mount Morris, Michigan*

An avid outdoorsman, Fred Harrington and his two sons enjoy hunting with bow and rifle, and also do lots of fishing. He has been interested in things artistic since youth, and is mostly self-taught as an engraver.

At seventeen he joined the Navy, where he served for four years as a quartermaster-signalman aboard the U.S.S. *Cassin Young*. During his free time aboard ship, he did pen and india ink drawings.

Mr. Harrington became interested in arms engraving in 1975 and did his first work on a gun in 1977. He made his first chisels out of hand files. Though he works as a tool salesman, he rarely lets a day go by without engraving — usually between noon and midnight.

Like many of today's engravers, Mr. Harrington does scrimshaw on firearms accessories, knives, and jewelry. He uses the chase method exclusively, engraving with only hammer and chisel.

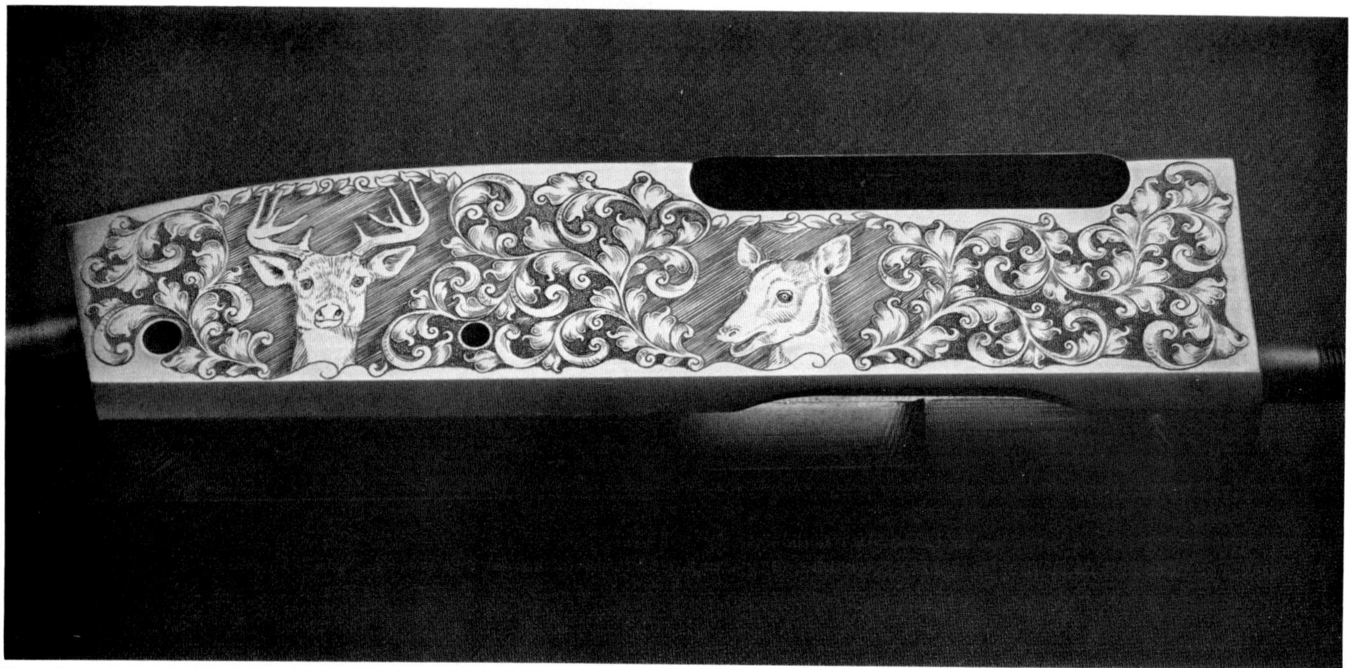

**These two views of a blued Remington Model 760 rifle action illustrate one of Mr. Harrington's latest engravings. Here he has combined bas-relief American style scroll with line-engraved game scenes. Note the stylized signature, F.A.H., under the word MODEL.** (Photo by Dave Strand)

**This stainless steel Smith & Wesson Model 67, belonging to the engraver, has been treated with high relief scrollwork having a stippled background. This is one of Mr. Harrington's earlier works.** (Photo by Dave Strand)

**Mr. Harrington's first effort at engraving a firearm was on this Marlin .30-30 caliber lever action rifle. The gun was presented to a young man on his sixteenth birthday.** (Photo by Dave Strand)

# Glen Hartliep

*Tucson, Arizona*

Glen Hartliep is the son and former apprentice of engraver Neil Hartliep, whose biographical sketch follows. Glen began doing gun work as a hobby when he was still in grade school. He received special recognition, with an A+ grade and his picture in the papers, at the age of twelve, when as a school project he made a Mannlicher-style stock by hand for the barreled action of a Mauser rifle. As a youth, he also began to learn the fundamentals of engraving.

After his four-year tour of duty in the Air Force, his career as an engraver started in earnest. He returned home from the Panama Canal Zone and began work with his father on their Bicentennial rifle project. For this undertaking he did all of the carving; he also engraved the scope rings and barrels of the fifty-four rifles.

In the spring of 1977, Glen Hartliep established his own shop in Tucson, Arizona, where he works today. Although engraving takes a great deal of his time, he still participates in his hobbies of flying, trap shooting, and music.

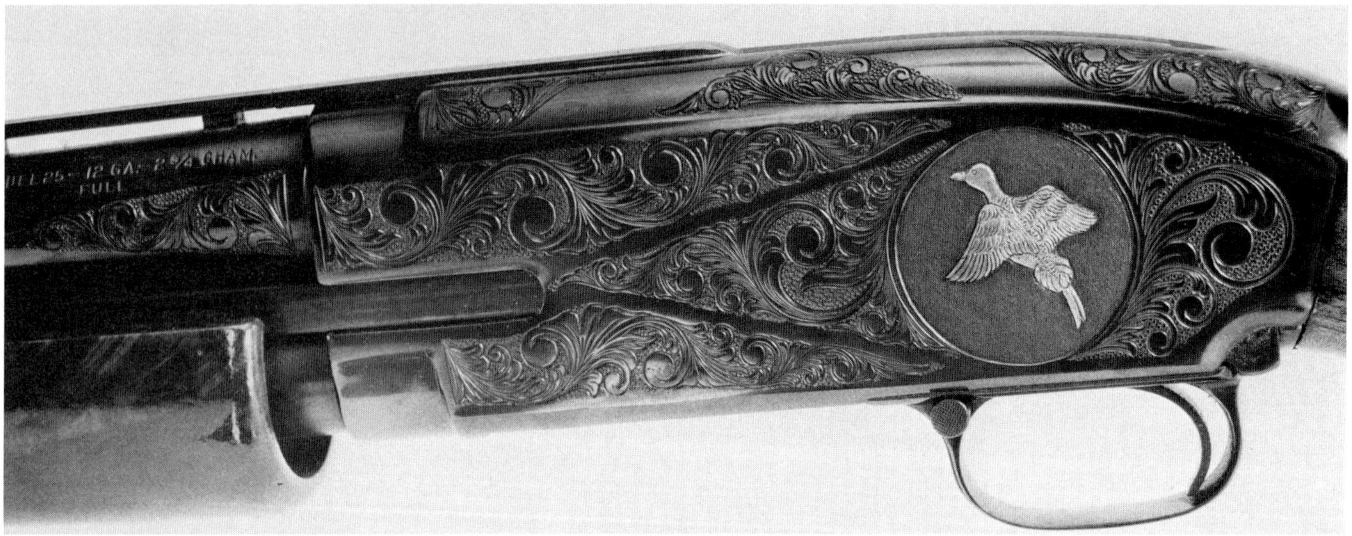

Glen Hartliep has upgraded this Model 25 Winchester with bas-relief American scroll and a raised gold inlay.

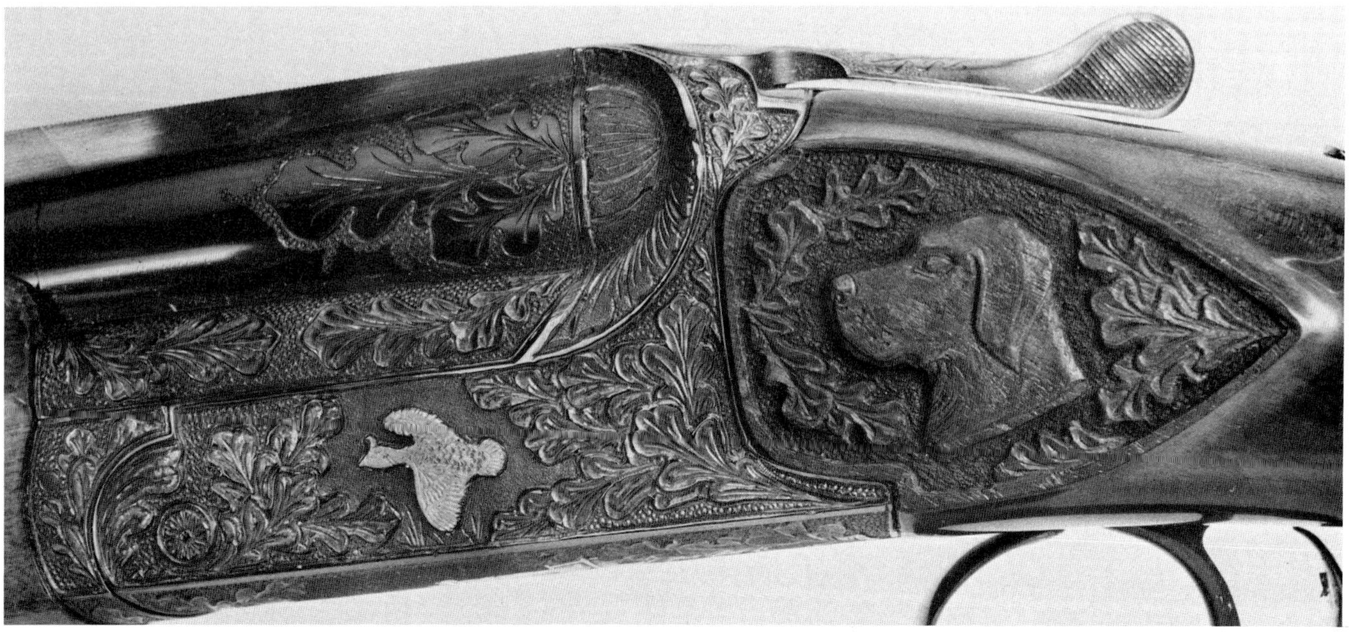

This Italian Perazzi single-barrel trap gun carries oak-leaf decoration and a gold-inlaid quail. The stock carving is also by Glen Hartliep.

# American Engravers: Neil Hartliep *63*

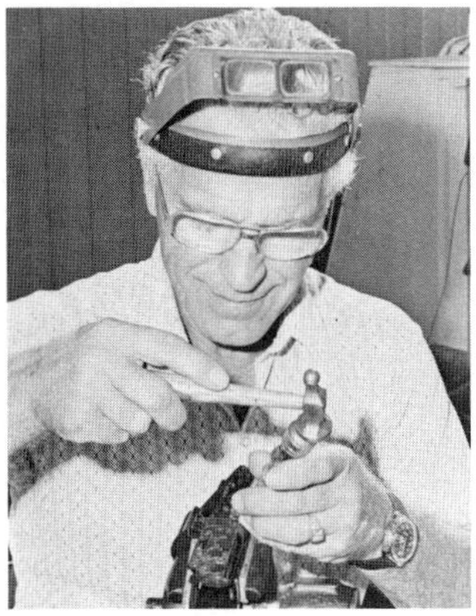

## Neil Hartliep

*Fairmont, Minnesota*

Raised on a farm in northwest Iowa, Neil Hartliep developed an interest in firearms at a very early age. At the age of nine, he used his .22 caliber Stevens "Crackshot" to eliminate farm pests and provide rabbits for the dinner table. By the age of ten he had saved, with the help of his grandfather, the $5.47 needed to buy his first shotgun, a 12 gauge "Long Tom" with a 42-inch barrel. In the more than fifty years that have since elapsed, Mr. Hartliep has not lost his interest in sporting guns or shooting.

During World War II Mr. Hartliep worked as a civilian employee of the 177th Engineers, which built a military base and airfield near Anchorage, Alaska. When the war ended, he went to work for the Kautzky Sporting Goods Co. at Fort Dodge, Iowa, where he learned both the retail business and gunsmithing. After examining some fine guns that belonged to the company's owner, Joe Kautzky, Sr., Mr. Hartliep became interested in engraving.

In 1948 Mr. Hartliep, who is entirely self-taught, began to teach himself the engraver's art while running his own sporting goods store in Cherokee, Iowa. Because there were no engravers near him to learn from and he could find no books on the subject, his progress was slow and difficult at first.

Another common problem among beginning self-taught engravers is the acquisition of gravers and chisels tough enough to cut gun steel. Mr. Hartliep made his first ones from triangular files.

In 1953 he sold his retail business and went to work as manager of the local Chamber of Commerce. This allowed him more time to practice engraving. Then in 1964 he began engraving as a full-time occupation.

In June 1979 Mr. Hartliep was the guest instructor at Trinidad State Junior College. He taught the basics of engraving to students in a gunsmithing curriculum co-sponsored by the National Rifle Association, of which he is a member.

He also belongs to the Amateur Trapshooting Association and Ducks Unlimited. Among his clients are entertainer and arms collector Buddy Hackett and William F. Cody, Jr., a grandson of the famous frontiersman.

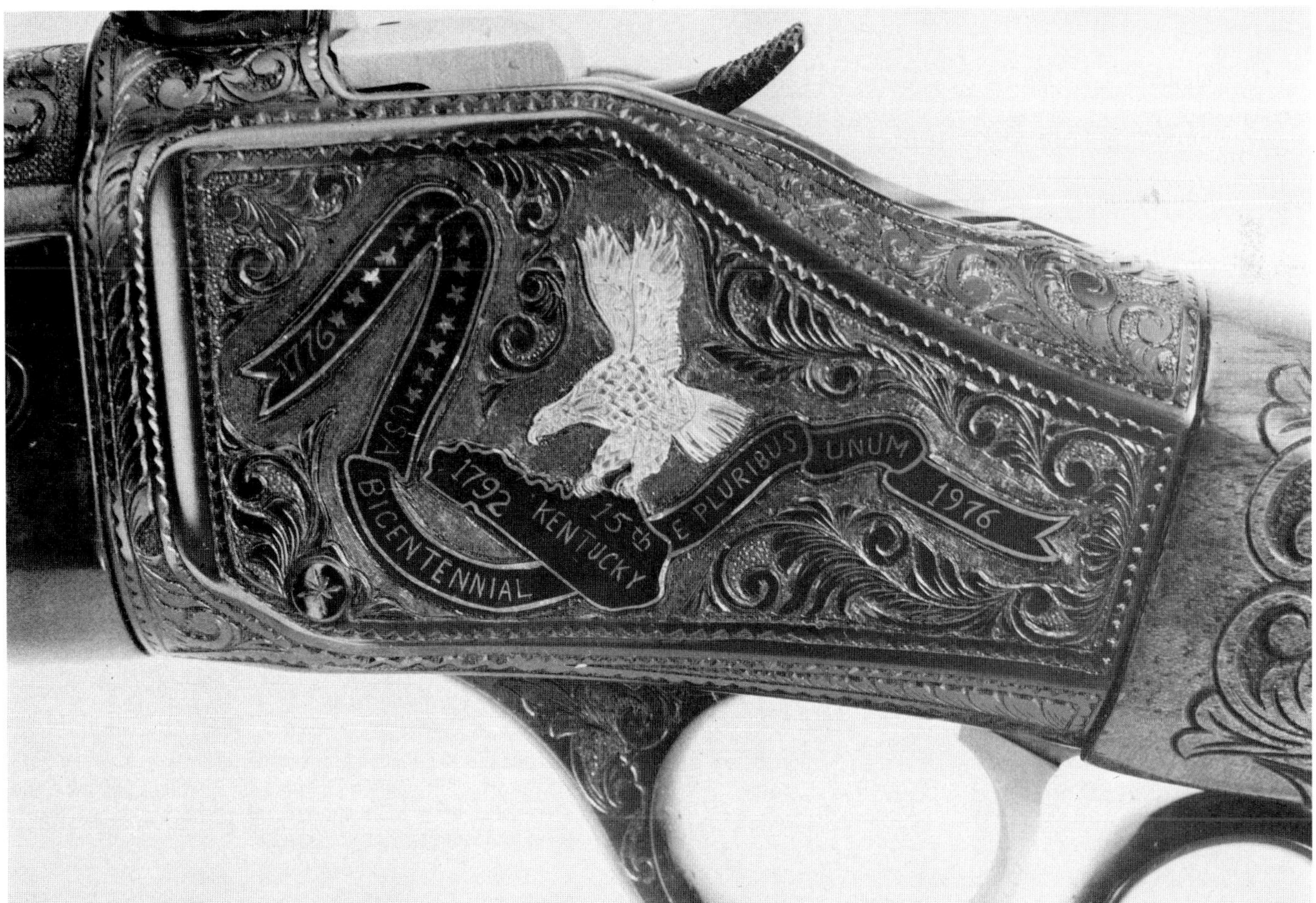


**Enlarged detail of one of the fifty Bicentennial Browning Model 78 rifles decorated for America's two hundredth birthday by Neil and Glen Hartliep. Most of the work shown here was done by Neil Hartliep. The wood carving and engraving on the barrel and scope mounts was done by his son Glen.**

The gray-finished receiver and blued barrel of this Ljutic trap gun have been profusely engraved in high relief by Neil Hartliep. This commission was completed in 1976.

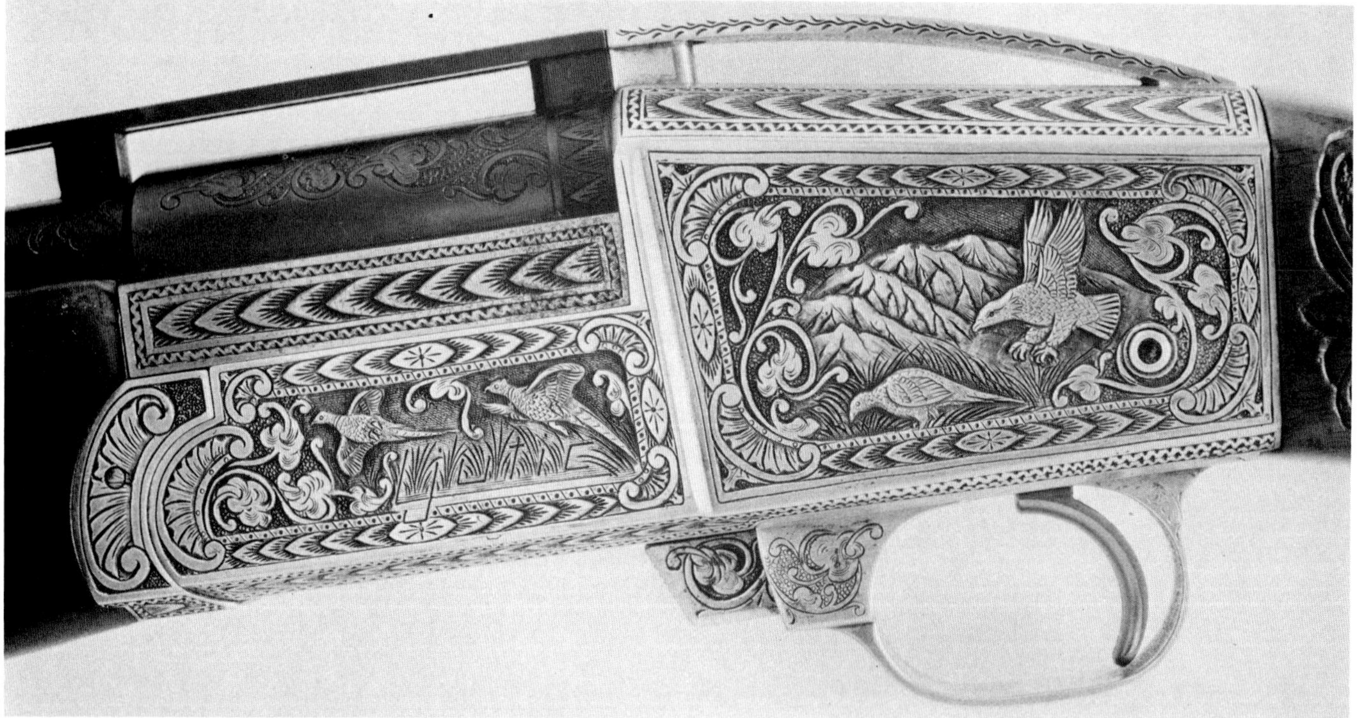

Engraved in 1976 by Neil Hartliep, this Ljutic Mono gun shows a variety of motifs, including birds chiseled in relief.

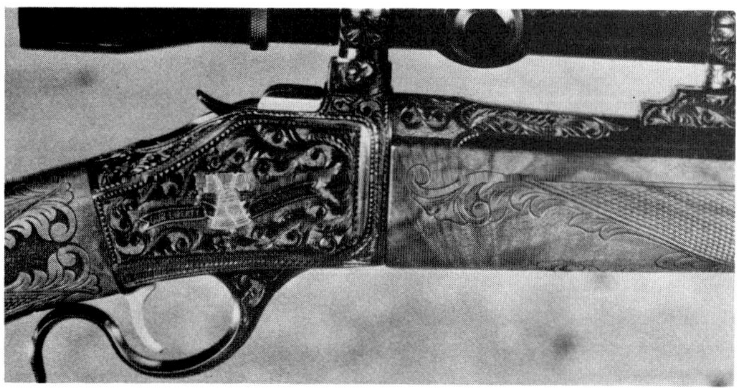

This Browning single shot rifle represents the combined efforts of father and son, Neil and Glen Hartliep. It is one of fifty such pieces made by them for the American Bicentennial. Neil, the elder Hartliep, engraved and inlaid the receiver, while Glen effected the barrel and scope mount engravings, as well as the carving of the wood.

The Krieghoff Model 32 shotgun pictured here shows Neil Hartliep's work in small, high relief scroll with flush gold line and figure inlays.

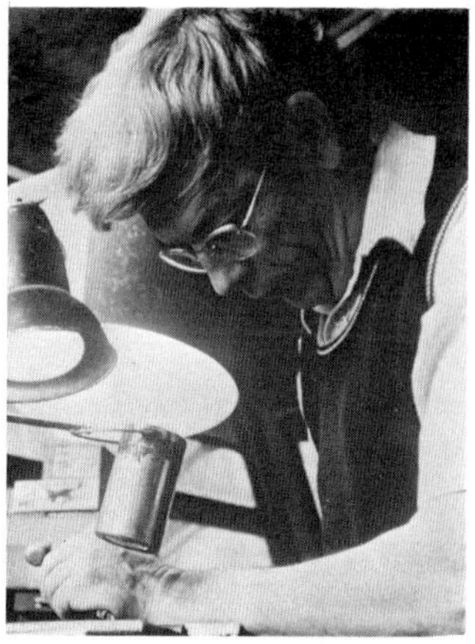

# Frank E. Hendricks

*San Antonio, Texas*

Born in McAllen, Texas in 1928, Frank E. Hendricks is a versatile engraver, proficient in over a dozen styles of art that are appropriate to arms decoration.

A ten-year veteran of the United States Air Force, Mr. Hendricks served in Germany, where he apprenticed under master engraver Kurt Rechnagel from 1954 until 1956. Although located in Frankfurt am Main, West Germany after World War II, Herr Rechnagel had engraved since 1905 for Sauer & Sohn in Suhl, East Germany.

Mr. Hendricks has been engraving professionally since 1961, when he set up his shop in San Antonio, Texas. His shop presently employs several helpers and engravers, and does all types of engraving, ranging from simple, less expensive patterns and production work to the most ornate and sophisticated art of museum quality.

In addition to engraving, the Hendricks shop does gun bluing and electroplating, and also some restoration of engraved arms.

For most of his cutting, Mr. Hendricks stands while using the chasing hammer and long wooden-handled chisels. For detailed work on scenes and figures, he cuts while seated. Most of his engraving is done under a suspended 3.5-power Bausch & Lomb engraver's lens. His vise is compact and very heavy, similar to a small milling machine or drill press vise.

**The triggerguard and floor plate shown here are fine examples of German scroll used by the late masters Kornbrath and Griebel; this technique is still in use by several masters in Germany and Austria. The scene is in low contoured relief; the sky is etched. Note the signature: ENGRAVING: F.E. HENDRICKS / S.A. TEX.**

**These two game birds were engraved on a Model 12 Winchester about 1964. Though they appear to be inlaid, they are actually relief cut in the steel and selectively frost-nickel-plated.**

This belt buckle was engraved in high relief by Mr. Hendricks in his spare time. The scene is from Charles M. Russell's painting, "The Buffalo Hunt."

German scroll is used here in combination with a bank note style game scene. Note the spear motif at the right and Hendricks's signature at the base of the tree.

Large American scroll, as popularized by nineteenth century engraver L. D. Nimschke, is used on this floor plate. The inner parts of the scroll have been relieved and matted with a circle punch. The charging elephant is an iron inlay.

This Browning shotgun carries a game scene and oak leaves cut in relief steel, with the acorns inlaid in different colors of gold. The sky on each side has been etched for a smooth flat finish. This is a proper use of etching in conjunction with hand engraving and should not be confused with etching done from patterns in mass production.

Mr. Hendricks engraved this portrait of a young boy about 1969. He has since been commissioned to do several other relief portraits.

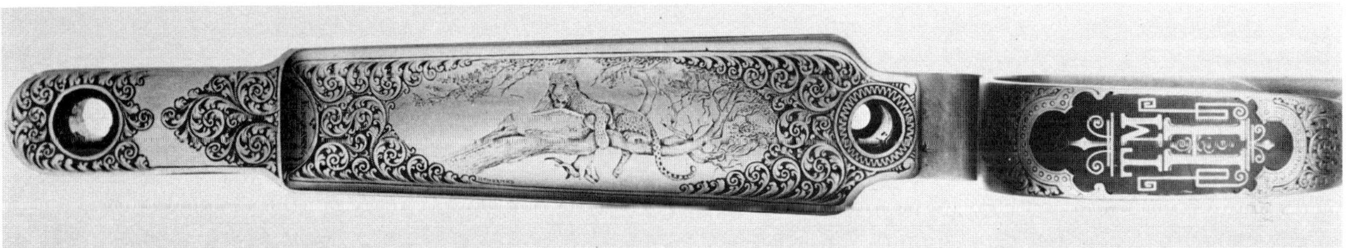

Mr. Hendricks calls the scene on this floor plate "Reclining Leopard With Prey." The initials on the triggerguard are fused gold inlay with an etched background for relief. The assembly has been frost-nickeled except for the background of the initials, which has been blued to contrast with the gold.

Engraved with a graceful and leafy style of scroll, the receiver sideplate and dust cover of this Model 1873 Winchester rifle are finished in frost nickel. The balance of the metal is blued with gold bands on the barrel.

# Benno L. Heune

*Santa Rosa, California*

After graduating from high school, Benno Heune went to work for the fire department of Modesto, California. In his free time he did wildlife drawings and hand painted neckties, which were popular in the 1930s.

During World War II Mr. Heune joined the Navy, where he served on a minesweeper patrolling near Okinawa, Korea, and, after VJ day, Japan. He continued his artwork in the Navy by painting insignias on ships and sea bags.

After the war he returned to work as a fire fighter. Eventually he met Jack Morris, who was making and engraving silver-mounted riding equipment such as bits, saddles, and other regalia for horsemen. For three years Mr. Morris taught Mr. Heune the applicable techniques of overlaying and engraving.

From 1955 until 1973 Mr. Heune owned and operated a sporting goods store in Bridgeport, California. Then he became a representative of the Weaver rifle scope firm until about 1976, when he began engraving full-time.

Mr. Heune's other interests include trap shooting, hunting, fishing, and flying. He is a member of the Scottish Rite of the Masonic Lodge and is also a Shriner.

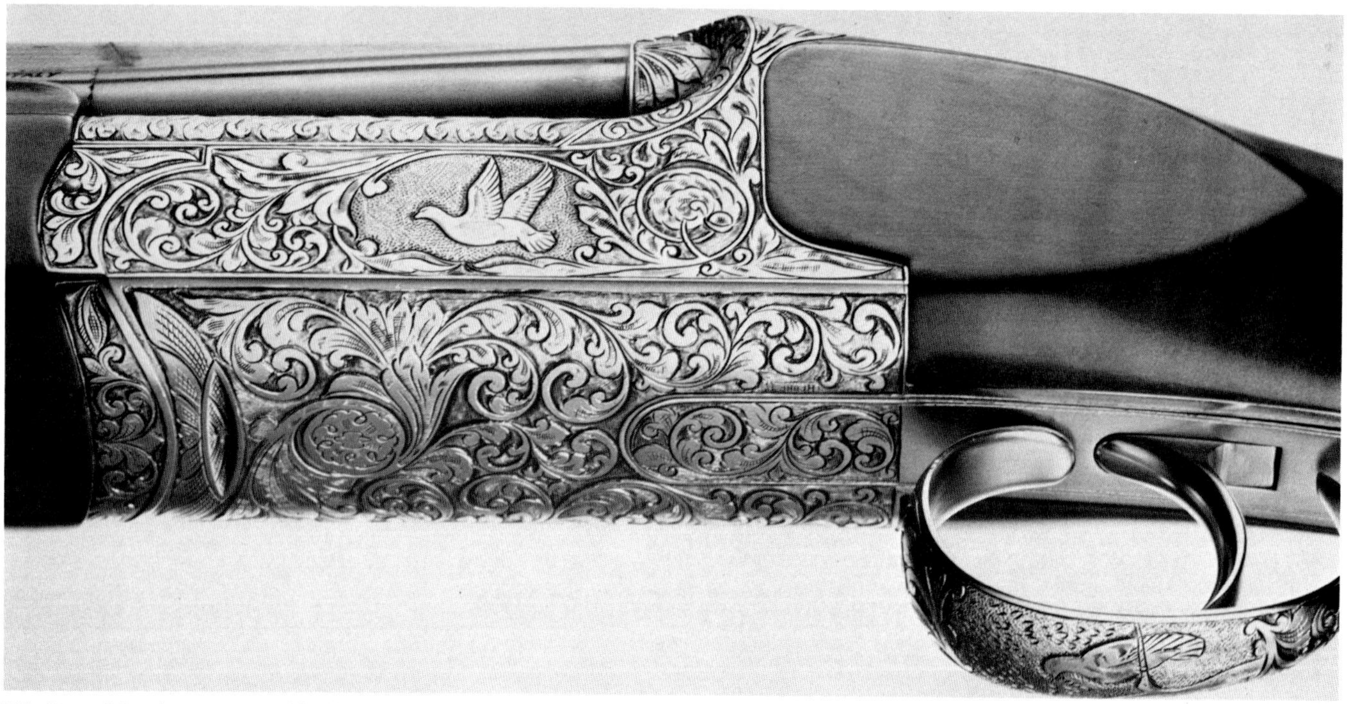

**This Perazzi has been engraved by Benno Heune with a floral scroll in medium relief. The background of the scroll has been treated with a liner.** (Collection of Dan Power. Photo by Ron Botier)

**Mr. Heune engraved the barreled action of this rifle with high relief scrollwork in 1978.** (Collection of Robert Thomas, Jr.)

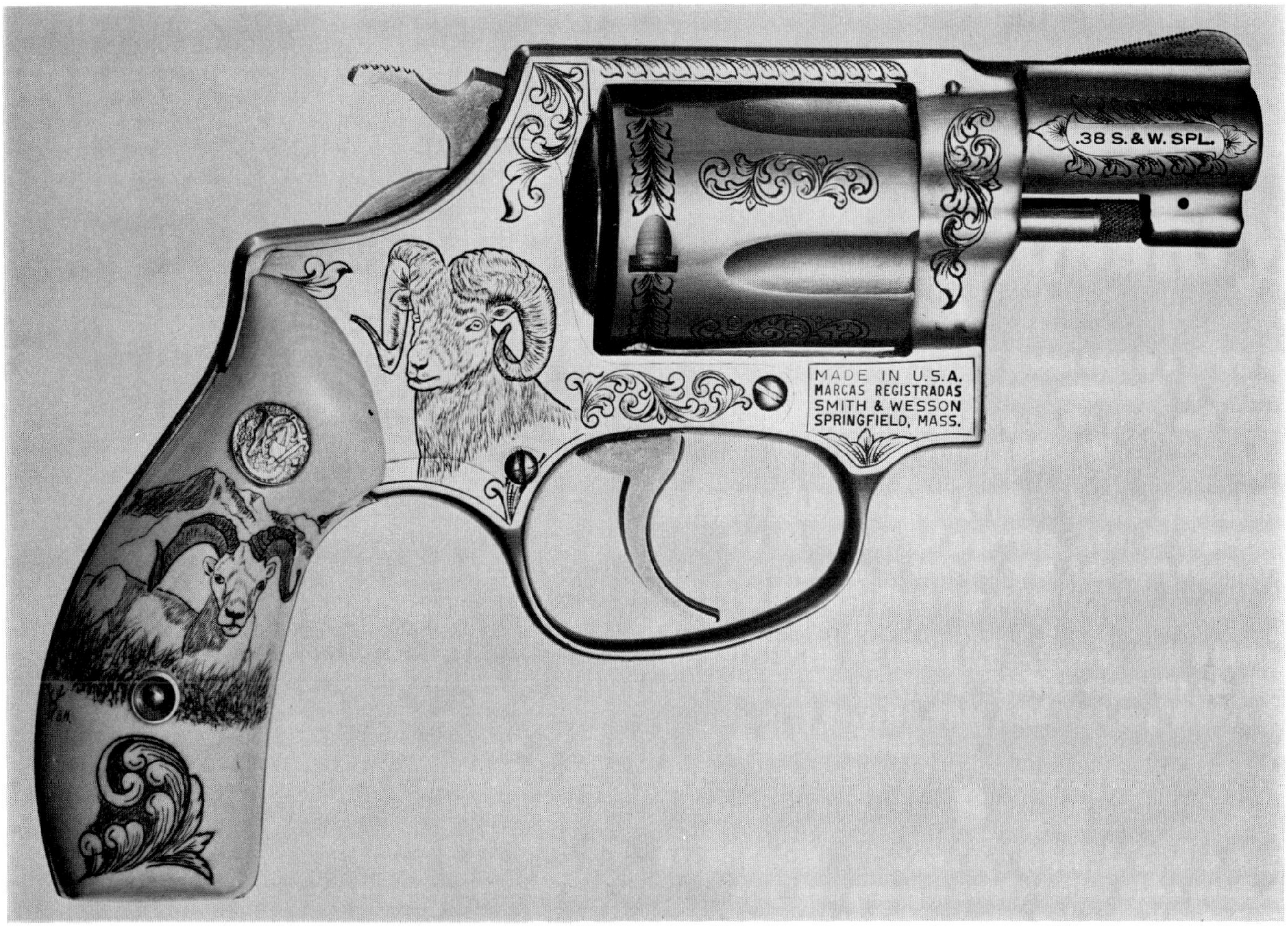


**With light scroll decoration and a line-engraved ram's head, this stainless steel Smith & Wesson also sports ivory grips scrimshawed by Benno Heune.** (Collection of William P. Croxton. Photo by Ron Botler)

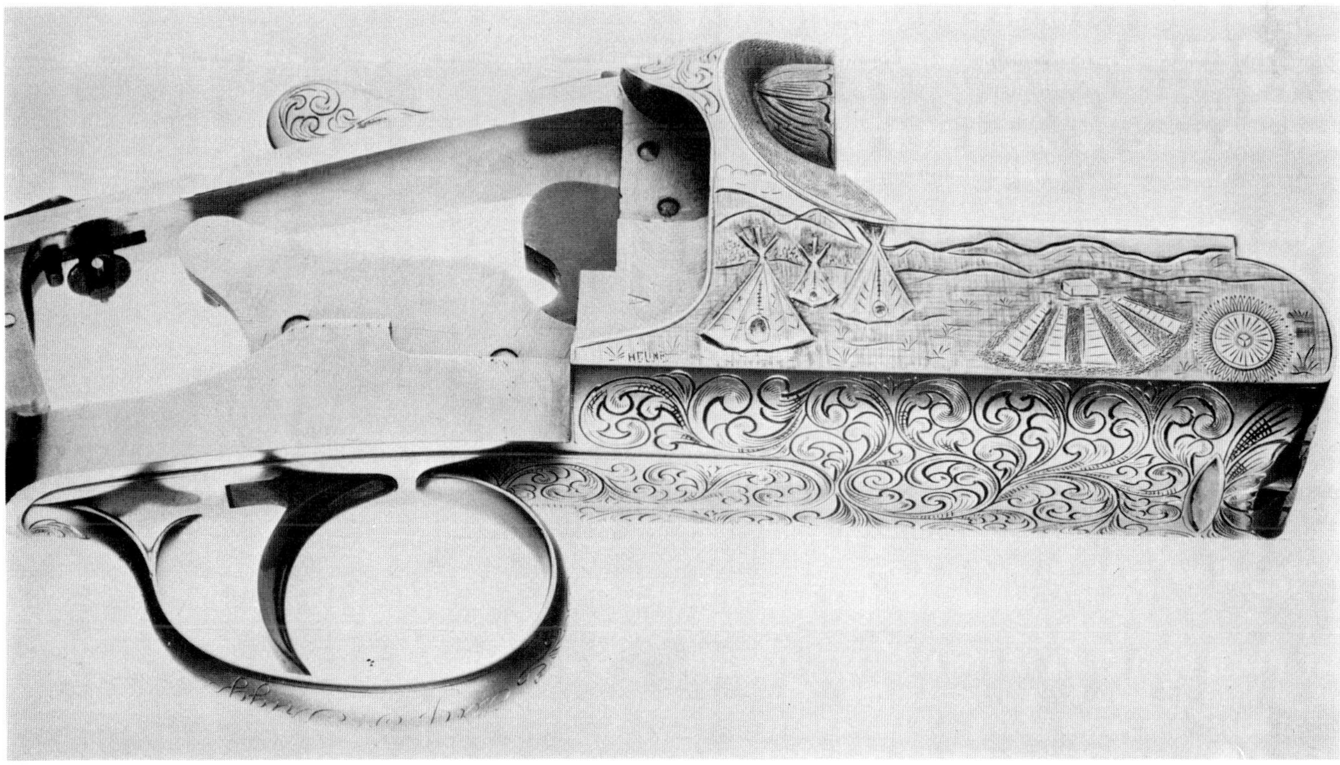

**Indian tepees with a trap shooting field to their right were engraved by Mr. Heune in relief on the receiver of this Perazzi trap gun. Note the signature in the lower left corner: HEUNE.** (Collection of Jay P. Griggs)

**Two views of a Winchester Model 42 shotgun in .410 gauge. Shown in the "white," this piece has been treated with large and medium size American scroll (sans background matting) and line engraving. The overall design is on the theme of Winchester's No. 5 pattern for slide action shotguns.** (Collection of Dayton Maddox)

# Tom Hicks

*Lubbock, Texas*

Tom Hicks is one of many engravers now practicing the trade in the state of Texas. He began his enterprise in 1976 and puts in about fifty hours per week at his bench.

He was first introduced to engraving by Frank Clark and subsequently was instructed by fellow Texan Ben Lane, who is featured later. Mr. Hicks does all of his engraving with the hammer and chisel; his work is held with a bench vise.

Mr. Hicks enjoys fishing and camping in his spare time. He also frequently displays his work at the gun shows in Amarillo, Fort Worth, and Lubbock, Texas.

**A view of a Colt Government model automatic in .45 caliber. The piece features borderless bas-relief scrollwork with circle-punched background and ivory grips. The finish is blued.** (Collection of Malcolm Tisdel)

**Another view of a Colt Government model automatic in .45 caliber. The piece features borderless bas-relief scrollwork with circle-punched background and ivory grips. The finish is blued.** (Collection of Malcolm Tisdel)

Mrs. Hiptmayer was born in Klagenfurt, Austria. Her training as an engraver began in 1957 when she entered the Ferlach Technical School of Firearms. After her graduation in 1961 she was employed as an engraver by the famous Austrian firm of Franz Sodia, which makes custom-built sporting arms in the town of Ferlach.

During her employment with Franz Sodia, she received a Meister Brief, a master's credential in arms engraving, from the Austrian government. Then after ten years Mrs. Hiptmayer, her husband Klaus (a stock maker), and their two children moved to Montreal, Canada. There Mr. and Mrs. Hiptmayer worked for Browning Arms of Canada.

In May 1972 the Hiptmayers opened their own business in custom arms, known as Austrian Gun Works Reg'd. of Eastman, Quebec. For her arms engraving Mrs. Hiptmayer uses only the traditional tools of chisel and burin while seated at the ball vise.

# Heidemarie Hiptmayer

*Kelowna, British Columbia, Canada*

**Still in the "white," this Model 21 Winchester skeet gun features medium relief German scroll and gold-inlaid figures. Note the rosette treatment of the hinge bolt.** (Photo by Studio R.C.)

**Another Model 12 Winchester engraved by Mrs. Hiptmayer. Featured are bank note-engraved Canada geese enclosed by an elegant border.**

True Germanic scrollwork in medium relief surrounds a regal looking elk sculpted in high relief by Mrs. Hiptmayer. Note the signature at the bottom of the floor plate: ENGRAVED / by / H. HIPTMAYER.

Possibly the most popular rifle used as a subject by American free-lance engravers is the Ruger No. 1 single shot, falling block rifle. This specimen in .30-06 caliber carries engraving by Mrs. Hiptmayer in medium relief Germanic scroll and high modeled relief game scenes. Pictured below the rifle is the profuse grip cap.

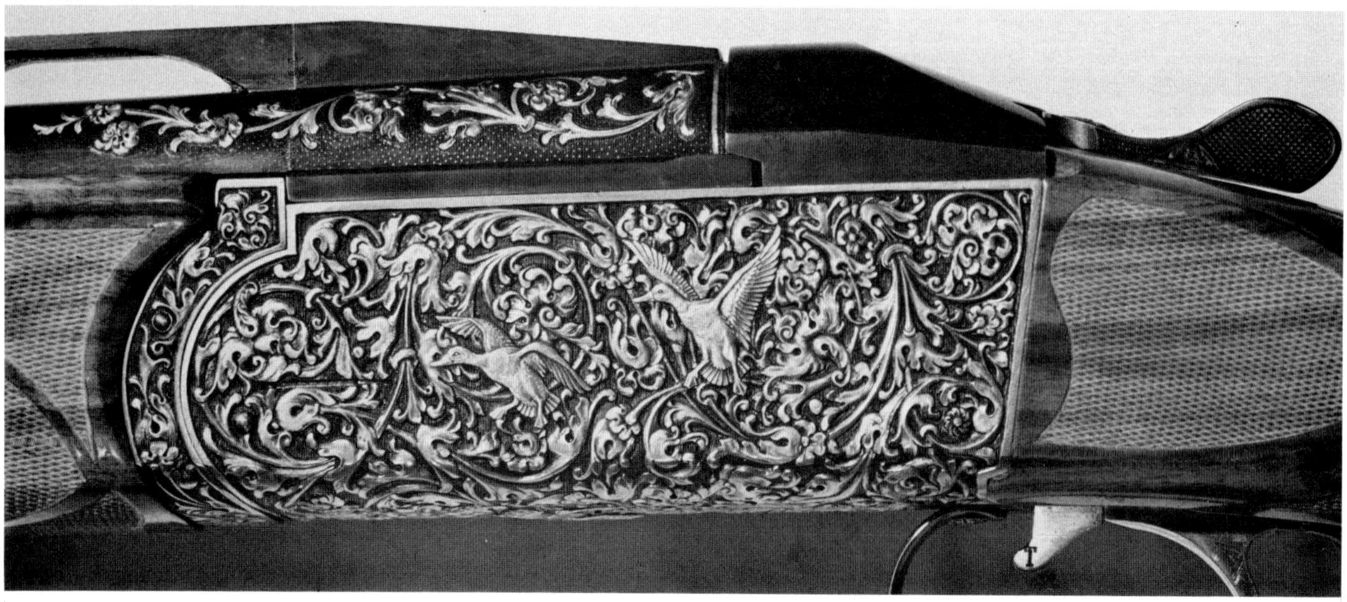

Mrs. Hiptmayer has sculpted relief vine and floral designs into this Krieghoff trap gun. The vines have been surmounted by water fowl of sculptured relief gold inlay. The barrel design is of raised sculptured gold also.

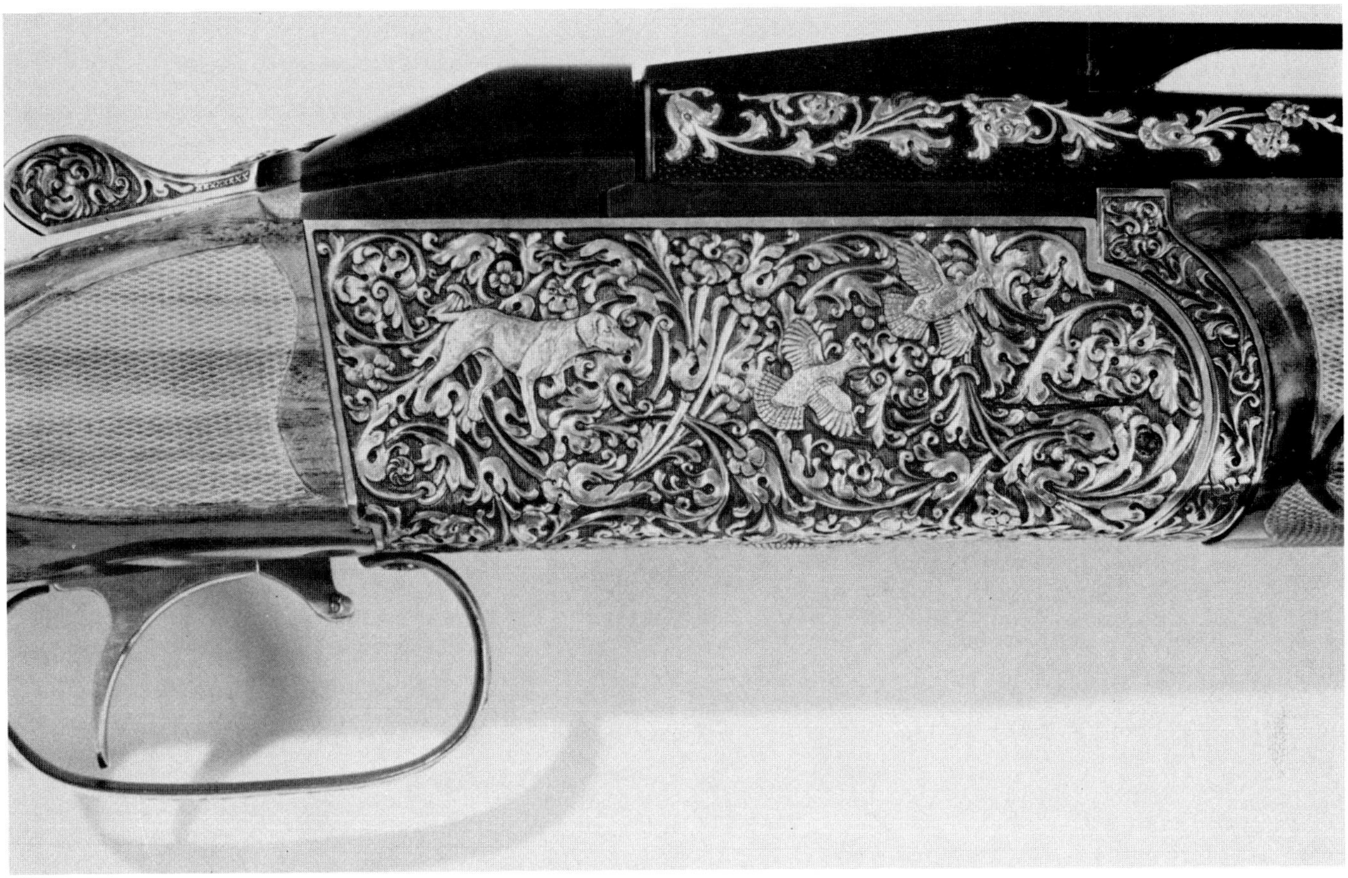

Sculpted with deep relief floral scroll and inlaid with gold figures, this Krieghoff shotgun is in the category of exhibition-grade work.

Shown here is a Ruger No. 1 single shot rifle engraved by Heidemarie Hiptmayer. The Ruger No. 1 is the most popular rifle for the work of contemporary American engravers. Mrs. Hiptmayer has decorated this piece with what most Americans call German scroll; however, Austrian-trained engravers like Mrs. Hiptmayer know this style as neuenglische Arabesken which translates as new English scrollwork.

**The ever popular Model 12 Winchester shotgun is pictured here with German scroll and high relief game scenes by Heidemarie Hiptmayer.** (Photo by Studio R.C.)

Born in Bloomington, Illinois and educated in wildlife biology at the University of Idaho, Mr. Huff is one of only a few engravers born in the United States and trained in Europe.

After college he studied under master engraver Rene Delcour at the Ecole de Fine Mechanique in Liege, Belgium. After studying for one year at the Belgian school he returned to the United States, where he went to work for the Colt Firearms Co.

Growing homesick for the western states, he left Colt after one year and set up shop in Montana. He works exclusively with hammer and chisel, and executes most traditional styles of decoration.

Mr. Huff is a veteran of the war in Viet Nam, where he served as a staff sergeant of artillery with the 101st Airborne Division. He has also worked as a blacksmith and enjoys horseback riding, rodeos, hunting, fly-fishing, backpacking, "white-water" boating, "big band" music, and ballroom dancing.

# Stephen E. Huff

*Missoula, Montana*

**This floor plate from a bolt action rifle by Al Biesen was engraved and inlaid by Steve Huff with high relief scroll and the likeness of a bighorn sheep in gold.** (Collection of George Douglas. Photo by Jon Schulman)

**Another floor plate embellished by Mr. Huff is from a rifle made by Jerry Fisher. Here a gold-inlaid elk appears to be wading in a stream and is surrounded by a cartouche.** (Collection of S. M. McCollum. Photo by Jon Schulman)

# Ken Hurst

*Lynchburg, Virginia*

Born in Greenville, North Carolina, Ken Hurst now lives and works in Lynchburg, Virginia in the foothills of the Blue Ridge Mountains. He was educated at East Carolina College and the Norfolk division of William and Mary, where he majored in structural engineering. During the conflict in Viet Nam he served with the Americal Division of the Army as a sergeant.

Mr. Hurst, who has been engraving about eighteen years, is mainly self-taught. His early interest was stimulated by what seemed to be the impossible complexities and many challenges of the art. He progressed from engraving inexpensive or wornout firearms to better work on finer guns. He developed his skill to the point where he was finally employed by the Colt Firearms Co.

In 1979 Mr. Hurst returned to free-lance engraving. Although he accepts work from private patrons, he also does considerable work for dealers and manufacturers. His commercial customers include the Wickliff Rifle Co., the New Sharps Rifle Co., and Southwest Shooter's Supply of Phoenix. His work for individuals has spread to such parts of the globe as Saudi Arabia, Brazil, Lebanon, Australia, and Panama.

The tools used by Mr. Hurst include the hammer and chisel as well as the NgraveR, a small hand-held engraving tool powered by a flex shaft motor. In addition to firearms and knives he engraves jewelry, nameplates, and ornamental art. At present, Mr. Hurst is training Bob Ligon, an apprentice he believes shows great promise.

**Detail of a Colt Single Action Army frame by Ken Hurst.**

**Detail of barrel and ejector housing, showing flush-inlaid maker's identification and relief scroll.**

**Detail of a Colt backstrap, showing a flush-inlaid spear intertwining with a scroll pattern by Ken Hurst.**

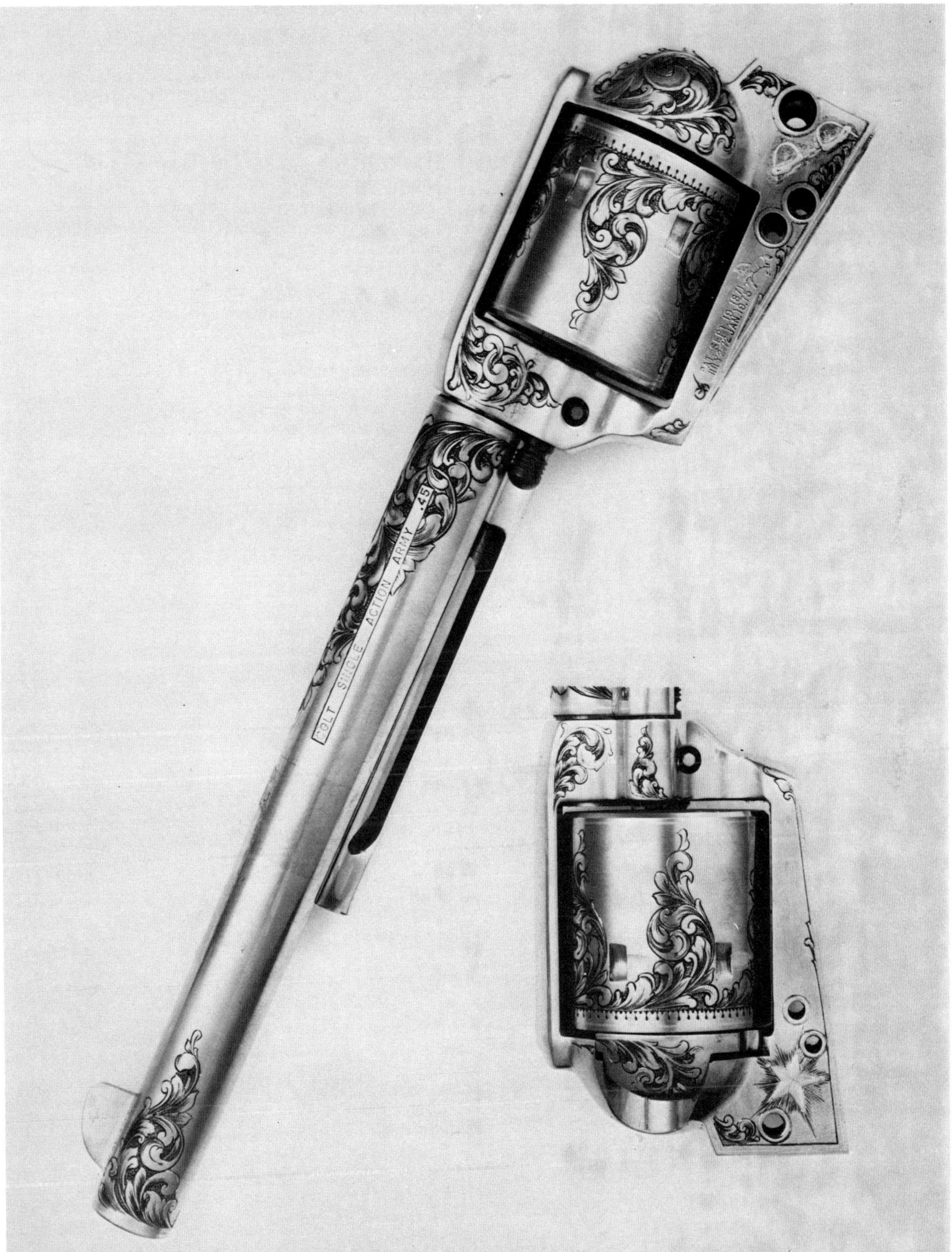


Shown here in the "white," this Colt has received flush wire inlays on the barrel, cylinder, and frame as well as raised inlays depicting the lone star of Texas and old-time handcuffs. Note this particular specimen has an unfluted cylinder.

# Ralph W. Ingle

*Rossville, Georgia*

Born in Chattanooga, Tennessee but now located in Rossville, Georgia, Mr. Ingle began engraving in the early 1960s. He served his apprenticeship under a master engraver, carving Coca-Cola bottle molds for a large firm in Chattanooga. Toward the end of his apprenticeship he began to engrave rings, knives, and cigarette lighters for his fellow employees.

Later he tried his hand at engraving a shotgun for a friend, who finally sold the gun back to Mr. Ingle. When showing the gun to others, he realized that he was very interested in engraving firearms, and eventually specialized in that work.

Aside from his background as an engraver, Mr. Ingle is trained in machine shop, welding, blueprint reading, and drafting. He enjoys hunting and fishing, and is a member of the National Rifle Association.

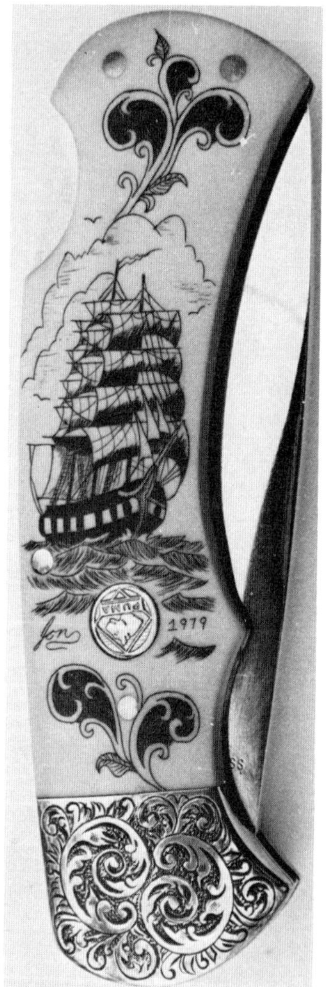


Mr. Ingle refers to his engraving on this Puma folding knife as "deep relief German floral." The scrimshander is Jon Poulakis.

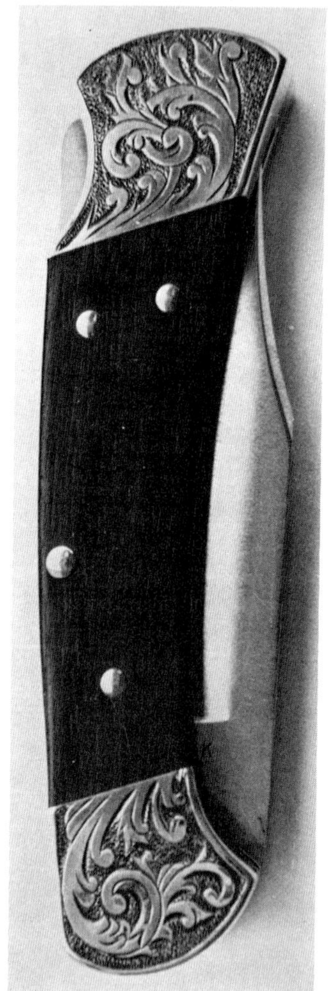

"Medium English scroll engraved in deep relief with black background" is how Mr. Ingle describes his work on this Buck Ranger knife.

This fully covered stainless steel Smith & Wesson has been relief engraved by Mr. Ingle in what he calls "small Old English" scroll. The background has been matted with a single point punch. Mr. Ingle also fitted stag grips to this piece.

# Bill Johns

*McAllen, Texas*

Bill Johns feels that he comes by his deep interest in firearms naturally. During his high school days he and his brother, Ray, worked on the family guns and tried their hands at bluing, plating, and stock making. Following service with the United States Navy from 1951 to 1955, Mr. Johns attended the University of Wyoming, where he received both his bachelor's and master's degrees in psychology and counseling. He worked for many years as a counselor at the Platte Valley High School in Saratoga, Wyoming. While in Saratoga he was active in the Lion's Club as well as the Masonic Lodge. He is a Past Master of Lodge #14 A.F.& A.M.

Mr. Johns began engraving while in the Navy, using tools that were a gift from an elderly retired jewelry engraver. He feels that his leisure time activities of horsemanship, hunting and fishing, and his interest in firearms have been assets to his arms engraving efforts.

In 1976 Mr. Johns moved from Wyoming to McAllen, Texas and has made engraving his full-time occupation. For many years he has attended gun and knife collector shows, where he often does some public engraving so that people may see how an engraver works.

Mr. Johns prefers to work in American style scrollwork. He also engraves in relief, carves ivory, and restores antiquated arms.

**Enlargement of a pair of Colt single actions shows the details of Mr. Johns's engraving. Note wiggle tool designs on the top strap of the upper gun and on the recoil shield of the lower gun.** (Collection of Doug Carlson)

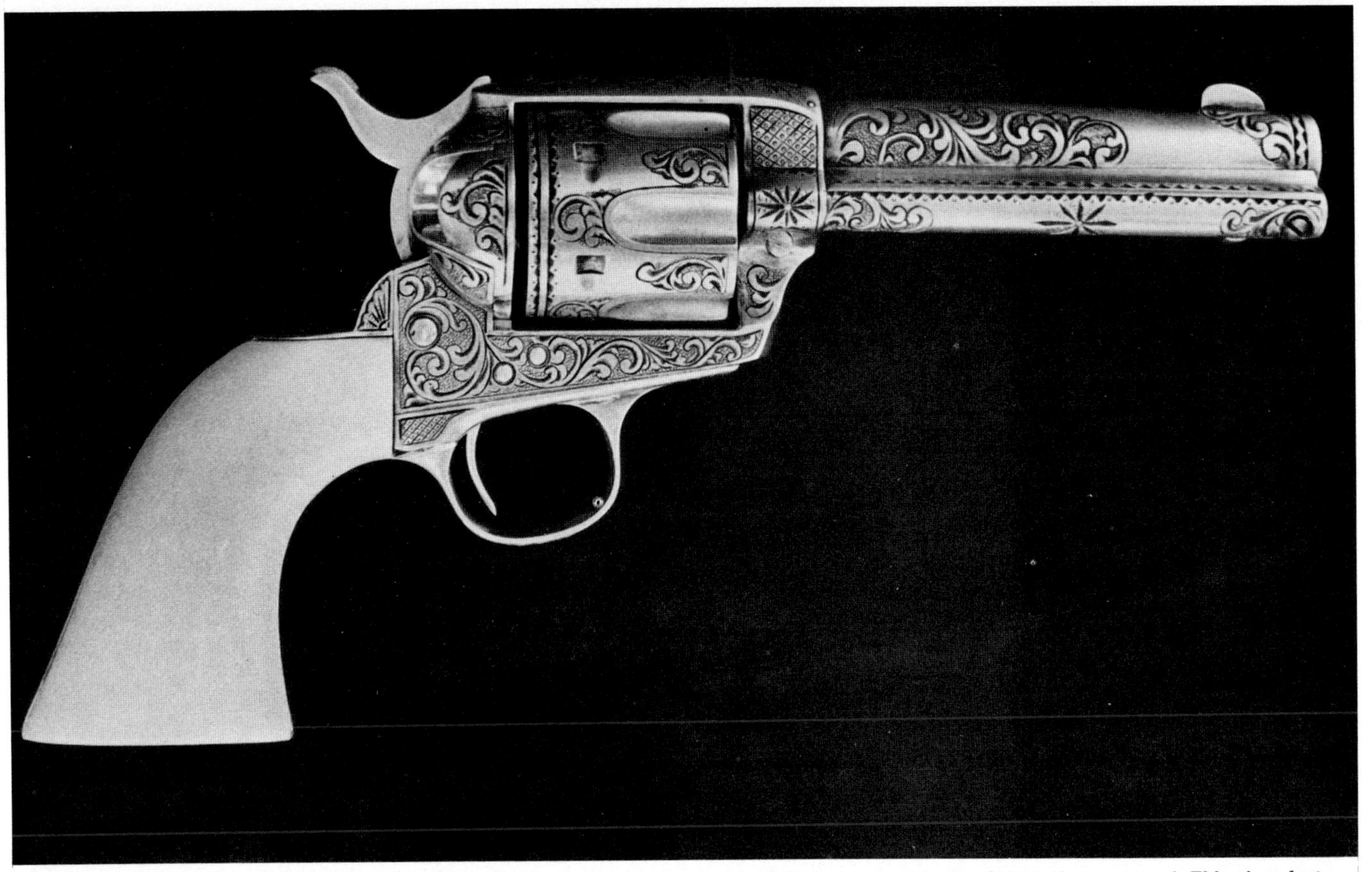

This Colt "Peacemaker" belonging to engraver Bill Johns is a representative sample of the many single action Colts he has engraved. This piece features antique silver plating and ivory grips.

**Sporting mother-of-pearl grips and antique nickel finish, this 4-inch barreled Python is decorated with floral scroll backgrounded with a liner.** (Collection of James Sinclair)

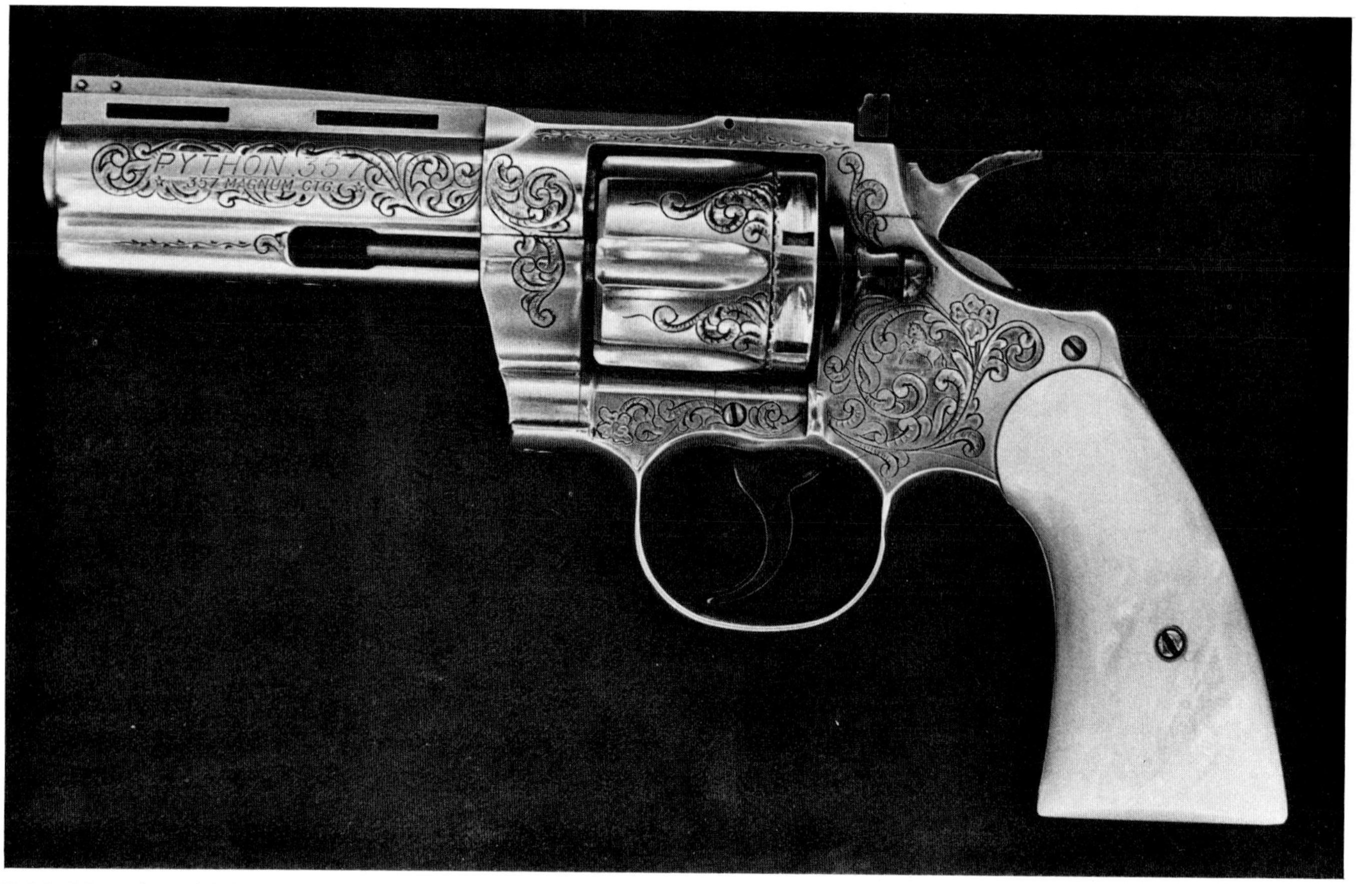


Finished in antique nickel and sporting genuine mother-of-pearl grips, this Colt Python was engraved by Mr. Johns for James Sinclair of Casper, Wyoming.

**Borderless scrollwork of about three-quarters coverage was used to embellish this Smith & Wesson Model 27 with a 5-inch barrel. The finish is antique nickel.** (Collection of James Sinclair)

At the age of sixteen, Robert Kain followed his two older brothers by going to work in the press room of a New Haven printing company, where their father had worked for many years. Then at eighteen he joined the Army, and after basic training served out his hitch as a corporal in Battery B of the 53rd Field Artillery at Masan, Korea.

Returning to the printing trade, he made a few extra dollars at his hobby of carving game scenes on the stocks of the local sportsmen's rifles. Eventually he left printing to be a professional carver and worked at a number of other jobs at the same time.

Mr. Kain regarded the engraving of gun steel as a great challenge and began to teach himself to engrave by the trial and error method. Progress was slow at first but as time went by his work improved until he developed his own unique style of scrollwork, which he has been using for many years.

In 1968 Mr. Kain received his first large contract, to engrave 2,000 Marlin Centennial rifles. The project took three years to complete, with the help of two apprentices.

His next large project focused on the battle of the Little Big Horn. He began by researching information at the National Archives and compiled the first documented list of men killed on that famous day in 1876. Then for each man, he engraved a Colt Single Action Army revolver with the trooper's name, rank, company, place of birth, date of enlistment, and circumstances of death. As a result of this research, he wrote a book entitled *The 7th and the Sioux, In the Valley of the Little Big Horn*.

Mr. Kain is also working on a Custer Memorial project for Harrington and Richardson Arms Co. In the same vein as the Colt revolver project, the Custer Memorial commission will be engraving on Trapdoor Springfield carbines.

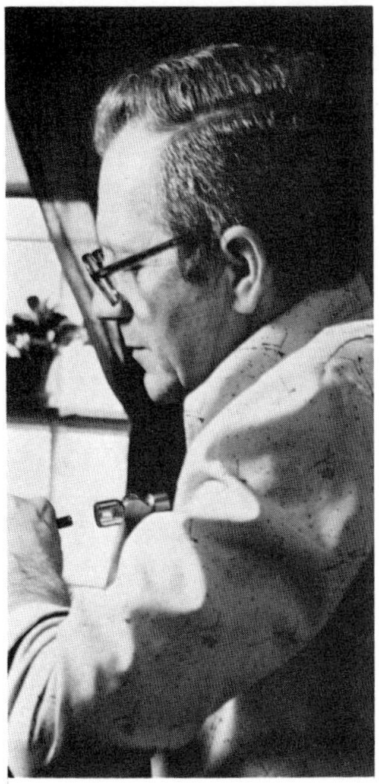

## Robert C. (Bob) Kain

*Canaan, Connecticut*

The heavy workload has made it necessary for Mr. Kain to form a corporation in partnership with David Townsend of Fall's Village, Connecticut, who is his business manager. Their intention is to develop the capacity of volume engraving in the unique Kain style. At present, Mr. Kain has one apprentice and hopes to add three more as time for instruction permits.

**Mr. Kain's hands are shown here applying fine detail with the burin to the 24 karat gold bust of William Lyman. The gun is a Ruger No. 1 rifle.** (Photo by Mike Brett)

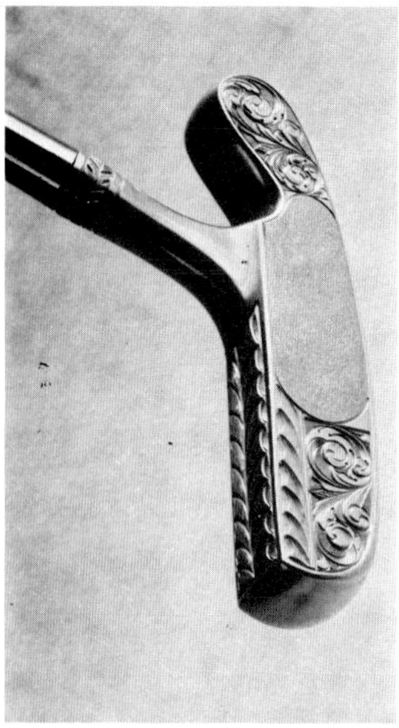

Two views of a golf putter engraved by Mr. Kain for a famous golfer. Note the "Bat Wing" motif along the top edge and upper border of the sides.

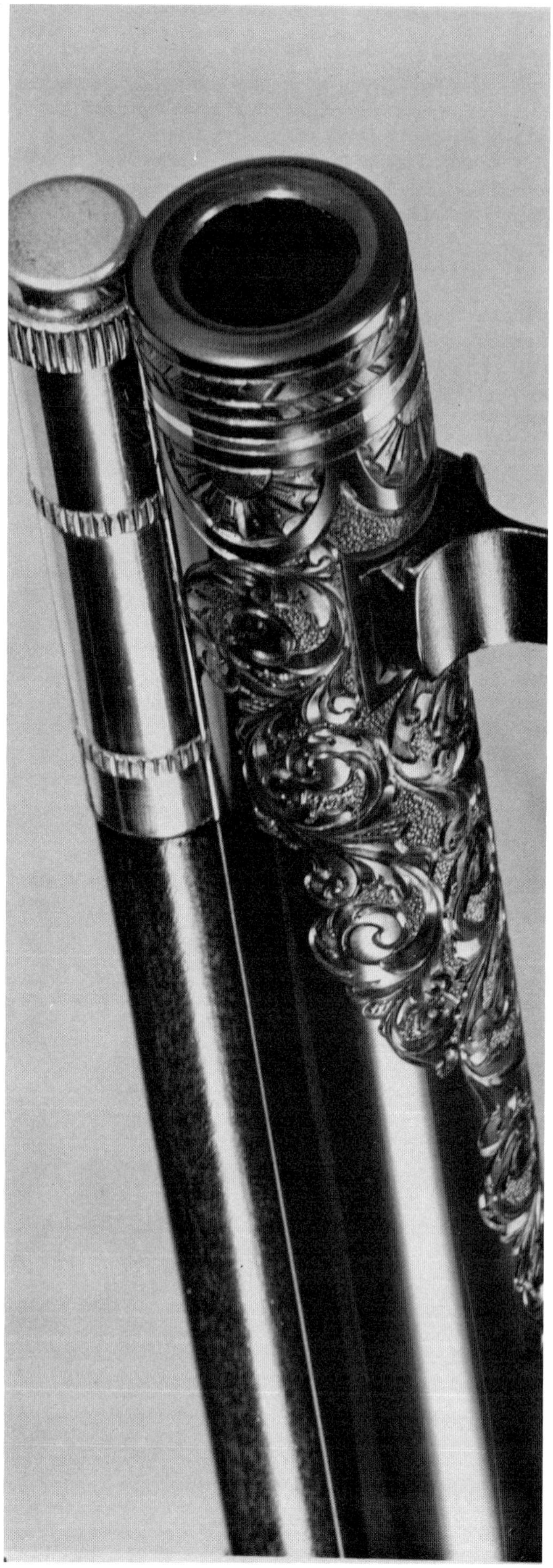

Detail enlargement of the muzzle of the Little Big Horn commemorative shows Mr. Kain's unique style of sculptured scroll with matted background.

# Tommy J. (T. J.) Kaye

*Beaumont, Texas*

A native Texan, Mr. Kaye was born and raised in Houston. During his formative years he was fascinated by the stories of the famous outdoor writers. At the age of fourteen, he read a story in *Outdoor Life* magazine by the late Jack O'Conner, and as a result he bought a Winchester Model 70 rifle, which is still used by his oldest son.

Mr. Kaye attended Lamar University in Beaumont, Texas, and graduated in 1961 with a bachelor's degree in commercial art. While at Lamar he met his future wife, Carolyn. Today they have two teenage sons.

After years of admiring fine arms at collector shows and in books, Mr. Kaye decided to learn engraving, the finishing touch to any fine gun. Like most American engravers he is self-taught. He feels that there are some advantages to learning in this way, though a long time is required to master the art. It took him about five years to develop his skills to do what he considers respectable work.

After eighteen years as a marketing representative with Texaco, he gave up his job to become a full-time engraver. Though he has worked on all types of arms, Mr. Kaye prefers to engrave rifles and shotguns. In his opinion the small number of American custom rifle builders are making the highest quality rifles in the world today. But in the area of shotguns, he feels that the English and Europeans are still the masters of the smooth bore gun.

Mr. Kaye intends that each commission produce a piece of fine art and for that reason never uses patterns of set designs. He does most of the metal finishing by hand in his own shop and uses only traditional tools to execute the engraving.

He is a life member of the National Rifle Association and the Houston and Texas Gun Collectors associations. Mr. Kaye believes that a finely engraved firearm should be used by its owner to be fully appreciated.

**A floor plate by Mr. Kaye in frost nickel finish. Note the engraver's signature on the left end of the plate:** T. J. KAYE.

**Shown here in the "white," this floor plate was later blued to contrast with the gold-inlaid bear.** (Collection of George Douglas)

**American scroll and relief African game scenes complement the French-grayed receiver of this Ruger No. 1 by T. J. Kaye.** (Collection of Joe D. Clayton)

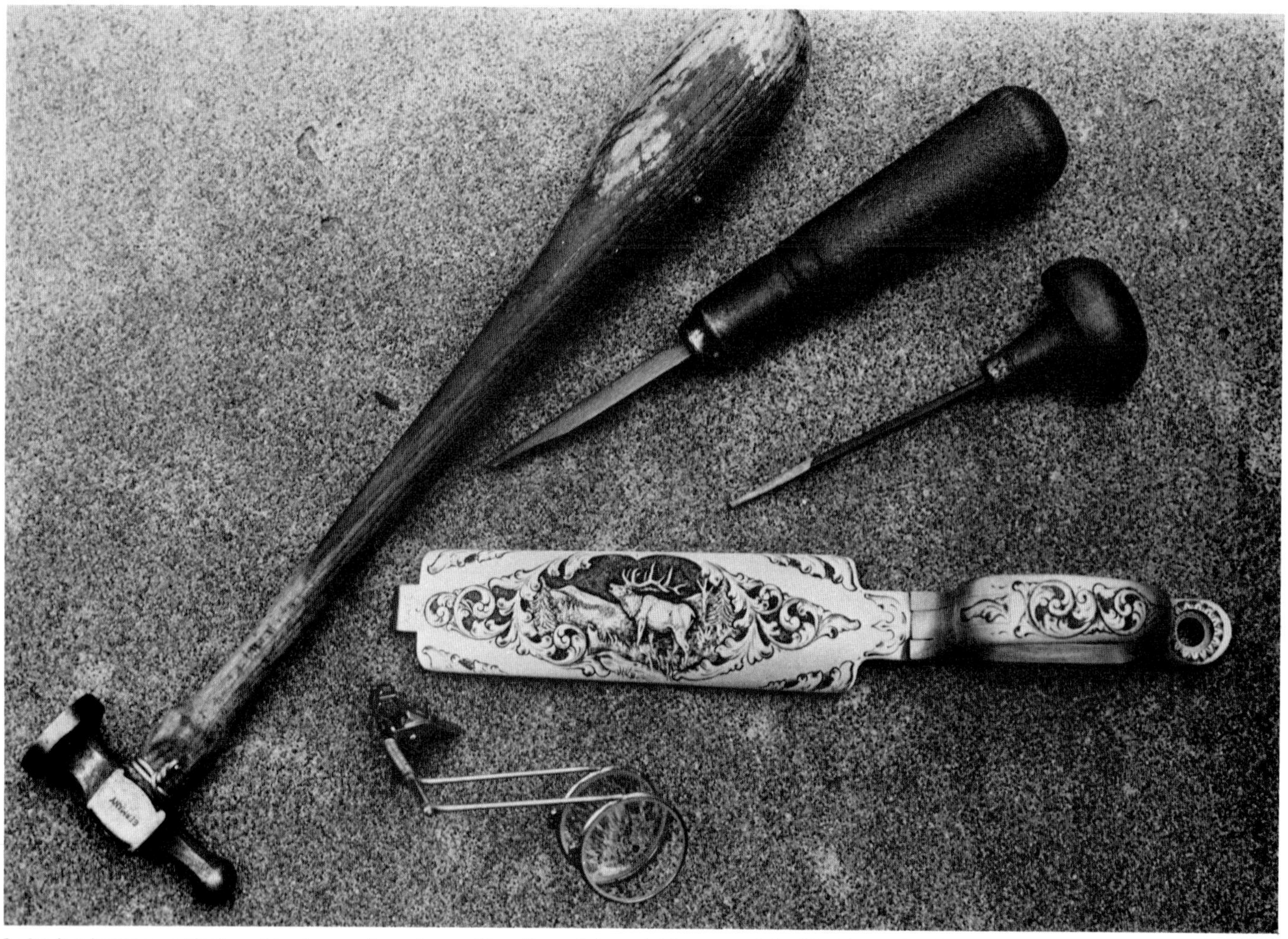

**A chasing hammer, chisel, and burin are shown here with a floor plate and triggerguard by Mr. Kaye. The jeweler's loupe pictured is meant to be attached to eyeglasses.** (Floor plate from the collection of Tom Skeuse)

**Bank note or bulino engraving was used by T. J. Kaye to decorate this steel grip cap, which has been finished with frost nickel.** (Collection of Tom Skeuse)

**American scroll and relief African game scenes complement the French-grayed receiver of this Ruger No. 1 by T. J. Kaye.** (Collection of Joe D. Clayton)

Educated in agriculture at Rutgers University, Mr. Kelly is also trained for work in machine shop techniques and art. His past also includes duty with the United States Army Air Corps as a lieutenant of ordnance. He served in the 8th Air Force for five years.

Like many of his American contemporaries he is mostly self-taught as an engraver, though he credits several arms engraver friends with help and useful advice. For his cutting, Mr. Kelly uses the hammer and chisel as well as the burin. To hold his work he uses both the ball vise, which is known by engravers as a "block," and the bench vise.

Mr. Kelly formerly engraved firearms, but since about 1976 he has concentrated on the custom knife field. At present he makes and also engraves a large variety of knives, both fixed and folding blade types.

He regularly displays his handmade knives and engraving at the Sahara Knife Show, which is held in conjunction with the prestigious semiannual Sahara Antique Arms Show at the Sahara Hotel in Las Vegas.

## Lance Kelly

*Decatur, Georgia*

**A Colt Single Action, extensively inlaid with flush gold spirals, is shown here with one of Lance Kelly's handmade folding knives, engraved with German style scroll.** (Photo by Weyer of Toledo)

These two views of a Colt Detective Special show that Mr. Kelly has used flush line inlay to form the master spiral lines of the scrolls, and then engraved the inner leaves in medium relief. (Photos by Weyer of Toledo)

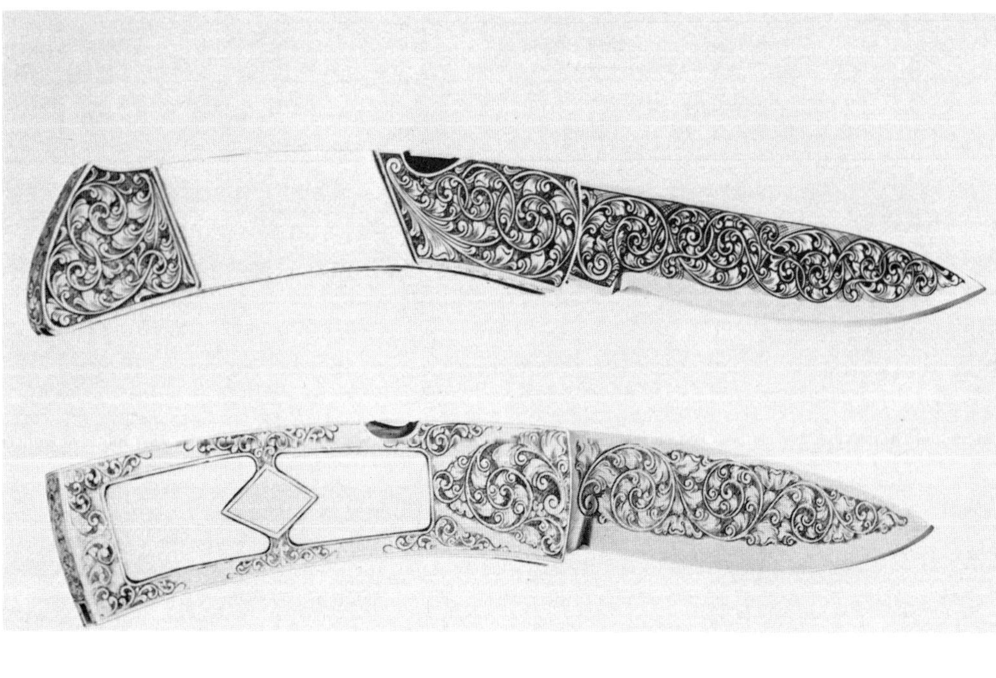

This selection of Lance Kelly's first model folders features German scrollwork and ivory handles. (Photo by Weyer of Toledo)

A relative newcomer to the field of arms engraving is Jim Kelso. Though born in California, Mr. Kelso has lived in the state of Washington for most of his twenty-nine years. He is fairly well-known on the West Coast for fashioning pearl-inlaid banjos, a trade that he pursued until fall 1977, when he began devoting all his time to arms embellishment.

His first encounter with arms engraving occurred when he traded one of his handmade banjos to engraver John Rohner for a Gravermeister tool and Magna Block vise. Then he acquired a book on arms engraving, which got him started with the basics. He has studied the work of other engravers as much as possible and feels that advice from professionals John Rohner, Robert Swartley, and Alvin White has been of great help in his development.

Mr. Kelso is convinced that it is important for an engraver to master the traditional forms of arms decoration, such as German, English, and American scrollwork; but he enjoys experimenting with less common forms of decoration. His goal is to engrave each arm as a one-of-a-kind piece incorporating his various skills, such as making inlaid

## Jim Kelso

*Baring, Washington*

or carved grips and handmade presentation boxes. One of his specialties is making fossil walrus or mammoth ivory grips and accessories.

Another of Mr. Kelso's skills is his training as a boat builder at the Seattle Community College. He also enjoys photography and building scale models with his young son.

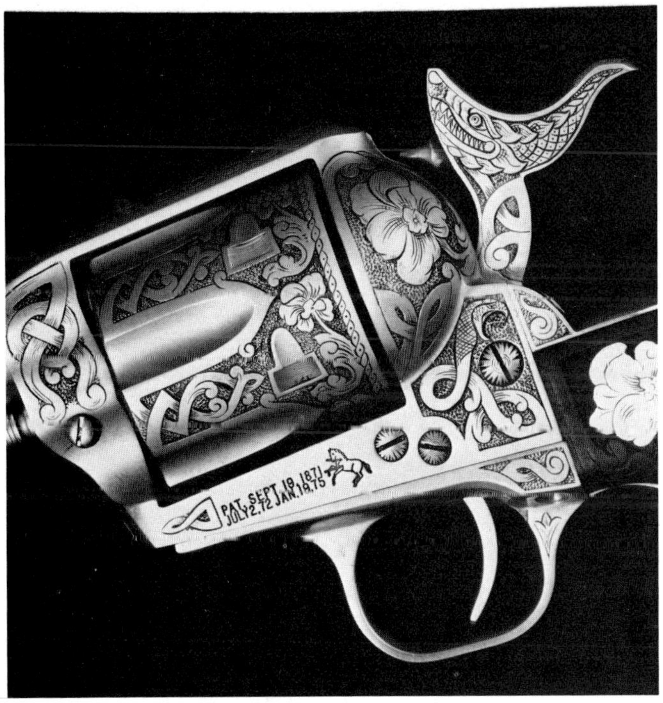

**Detail enlargement of "Celtic" style Colt, showing the wolf- or dragon-head hammer. Note the design forming a snake below the recoil shield.** (Collection of Dr. Lester Mittlestaedt. Photo by Mustafa Bilal)

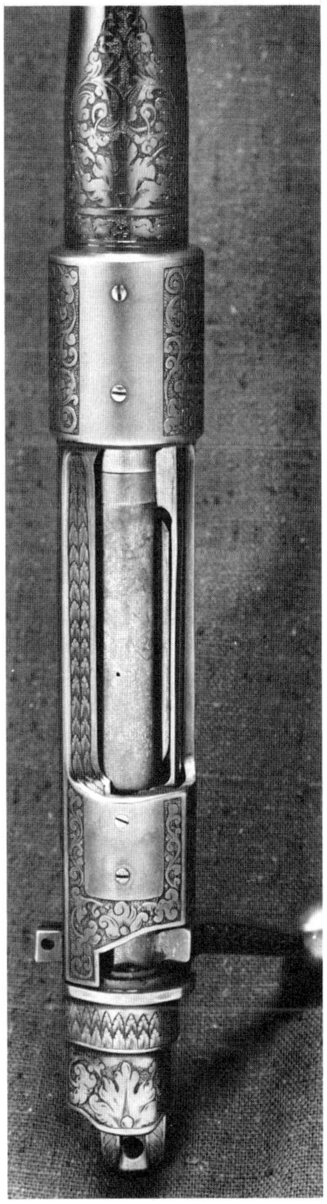

**This F. N. Mauser barreled action engraved by Jim Kelso features German leaf scroll with wheat chaff bordering. The receiver is dove gray with blued barrel in 7x57mm.** (Collection of Stuart Williams. Photo by Mustafa Bilal)

"Celtic" style designs is how Mr. Kelso describes the unique decoration on this Colt. The ebony grips were checkered and inlaid with mother-of-pearl by Mr. Kelso. Frost nickel finish is by Frank Hendricks. (Collection of Dr. Lester Mittlestaedt. Photo by Mustafa Bilal)

Work was still in progress on this .38-40 Colt Bisley when this photograph was taken. Barely visible are the gold bands on the cylinder and barrel. The elephant ivory grips are also by Jim Kelso. An inscription on the backstrap of this revolver, decorated to honor Chief Joseph, reads: I WILL FIGHT NO MORE FOREVER. (Collection of Dr. Lester Mittlestaedt)

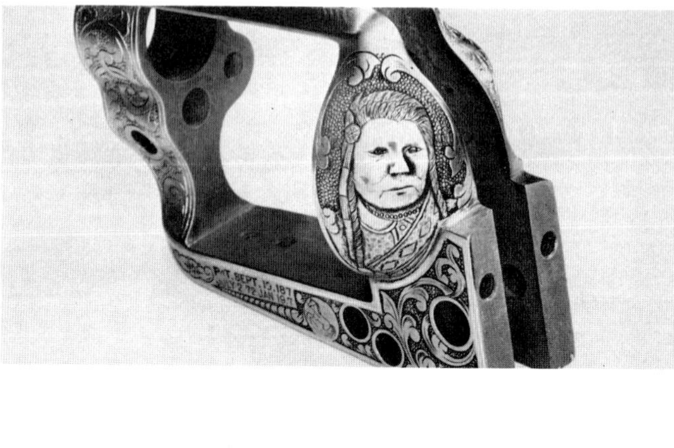

Enlarged detail of this Colt frame shows relief bust of Chief Joseph and the flush-inlaid arrow motif on the topstrap. (Collection of Dr. Lester Mittlestaedt)

A graduate of both Thackers Academy of Engraving in Pittsburgh, Pennsylvania and the Cummins Art School of Des Moines, Iowa, Mr. Koevenig has a solid background for his work as an arms engraver. His training also includes an apprenticeship with silversmith Tom Friet of Scenic South Dakota.

Mr. Koevenig has held other jobs that round out his skills, including work as a carpenter for the Bureau of Reclamation and chief of maintenance for the National Park Service, as well as a two-year stint in the Army Air Corps in the 1940s.

At present Mr. Koevenig works about thirty hours per week at his bench, where he executes oak-leaf decoration as well as English and American scrollwork. He frequently uses floral decoration in conjunction with relief figures, inlays, and overlays. Scrimshaw and acid etching are also in his repertoire. In addition he frequently decorates knives, jewelry, and buckles.

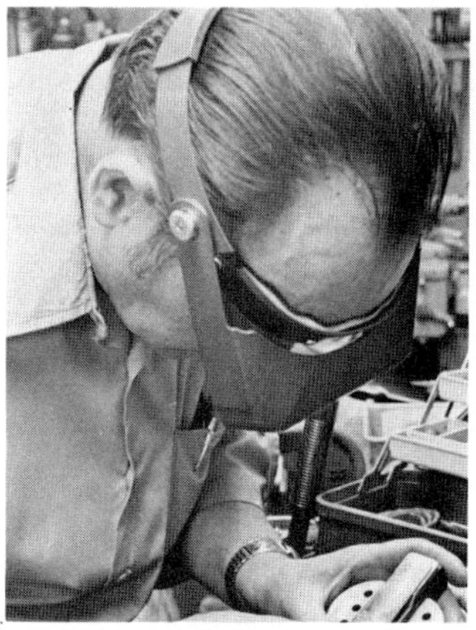

# E. J. Koevenig

*Keystone, South Dakota*

An advocate of fraternalism, Mr. Koevenig is a member of both the Benevolent and Protective Order of Elks and the Free Masons. He supports the formation of an American organization of engravers.

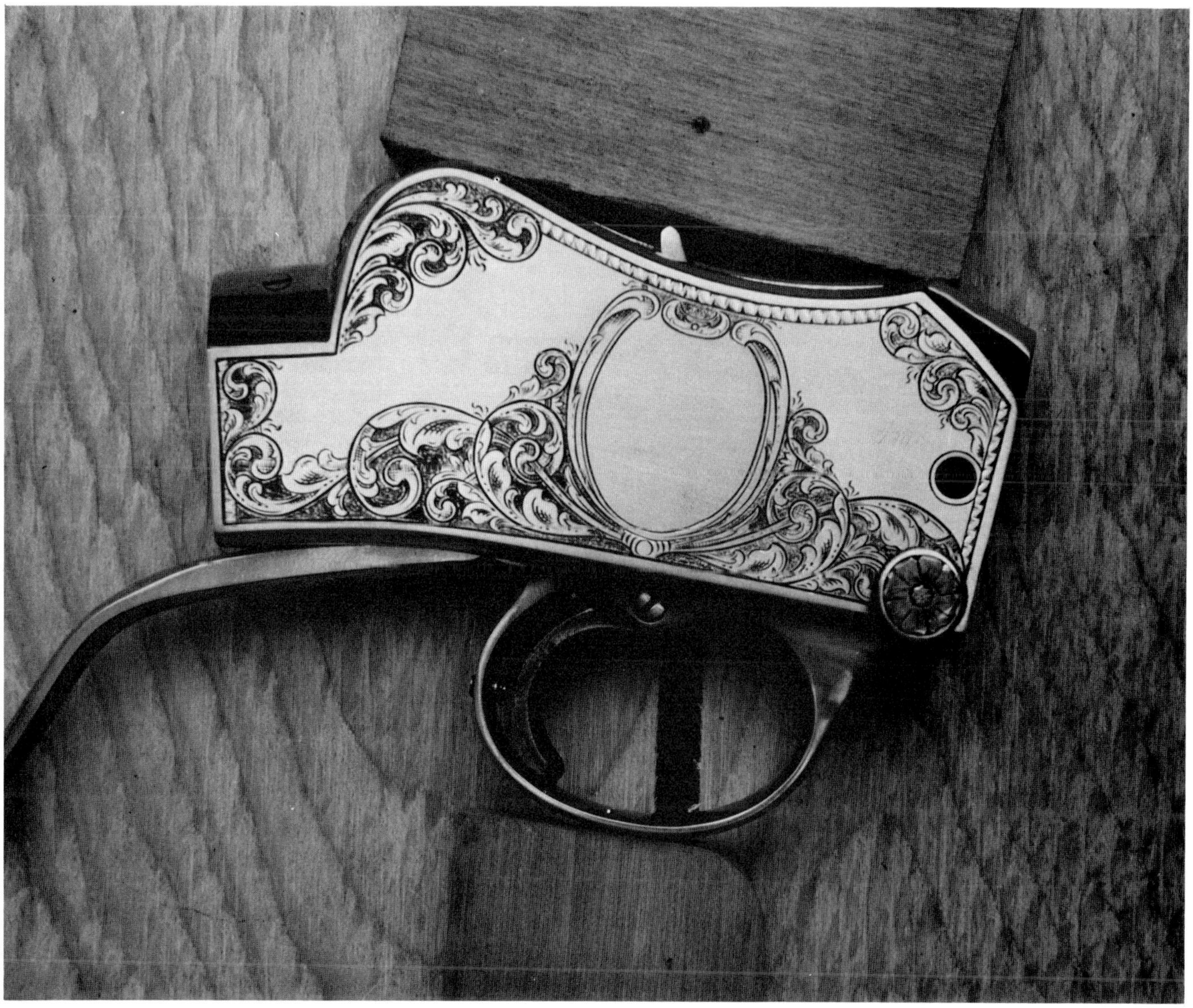

**Mr. Koevenig has engraved the receiver of this Martini single shot rifle with scrollwork and a central cartouche, so that an inlay can be done for the owner at a later time.**
(Collection of Bernie Larson)

# Walter Kolouch

*McMinnville, Oregon*

Though born in Vienna, Austria, Walter Kolouch (pronounced Ku'lock) has been a long-time resident of the United States. Though best known as an engraver, he is also qualified as a gunsmith and stock maker.

Mr. Kolouch's training began in Europe, where he worked for the firm of Johann Springers Erben, a custom gun maker in Vienna. This was his first job as a gunsmith/engraver after he graduated from high school. Later he trained at the "Gewerbe Schule" of Vienna and the "Fach Schule fur Handfeuerwaffen" at Ferlach, Austria. He also worked with master engraver Herr Schildbach of Mainz, Germany, as well as for other masters in Sweden and Switzerland.

Mr. Kolouch has engraved for several arms makers in both the United States and Europe, including Dan Wesson, High Standard, and O. F. Mossburg. For the latter firm he engraved a presentation rifle for Leonid I. Brezhnev. He also engraved a rifle commemorating the 350th anniversary of the Husqvarna Arms Co. of Sweden. This rifle was presented to HRH Karl Gustav XVI.

Though Mr. Kolouch does work for private individuals, a good portion of his engraving is done for dealers of fine sporting arms such as the world-renowned Jaqua's Sporting Goods firm of Findlay, Ohio and Southwest Shooters Supply of Phoenix, Arizona.

Mr. Kolouch is well versed at engraving all styles appropriate to arms decoration, including oak leaf, Germanic, English and American scrolls and floral. He works only with traditional hand tools and incorporates gold-inlaid figures into most of his commissions.

**Enlarged detail of Mr. Kolouch's bas-relief German scroll and inlaid game scene on another Krieghoff trap gun. Note the farm scene with silo in the background.**

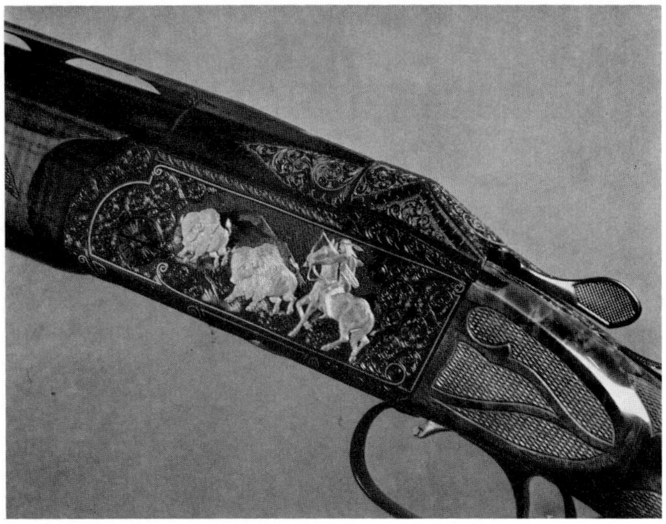

**Walter Kolouch has engraved this Krieghoff Model 32 Vandalia trap gun with German scroll and an inlaid scene of an American Indian hunting buffalo. During his career Mr. Kolouch has been called upon many times to depict scenes of western America on competition shotguns.** (Photo by Markow, courtesy of Southwest Shooters Supply)

**Enlarged detail of an inlaid scene by Walter Kolouch. The eagle seen here has been used on a variety of guns by Mr. Kolouch.**

**Ever popular with trapshooters, the Ljutic single-barrel shotgun shown here is engraved with bas-relief German scroll, backgrounded with the circle punch. The geese are raised gold.** (Courtesy of Bill Jaqua)

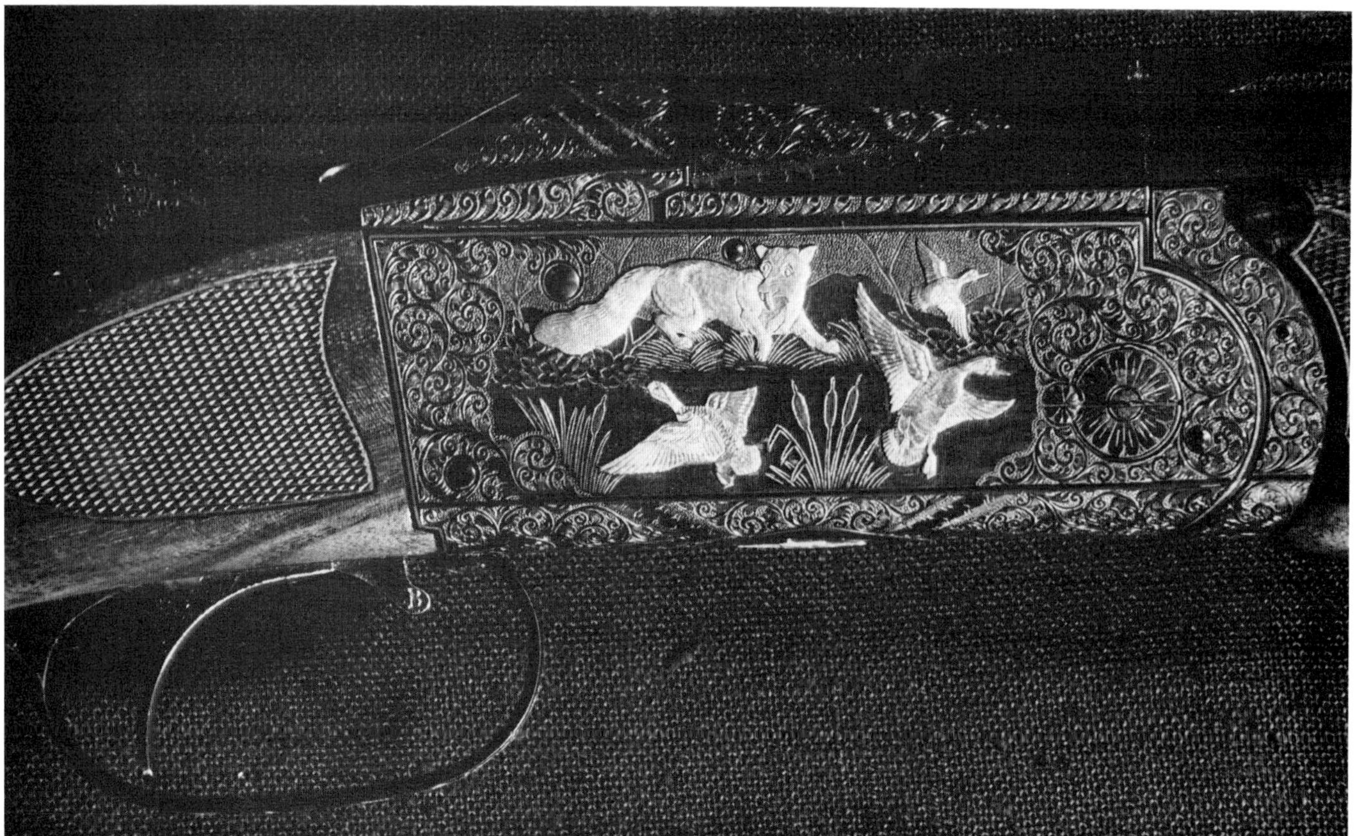

**Extensive coverage of Germanic scrollwork and gold inlays are regular features of arms decorated by engraver Walter Kolouch. This Krieghoff depicts scenes of predators and game birds in raised gold.** (Courtesy of Bill Jaqua)

Another of the many Krieghoffs engraved during Mr. Kolouch's long career, this one with an American Indian scene.

This small gauge Parker side-by-side features gold-inlaid doves on a grayed receiver.

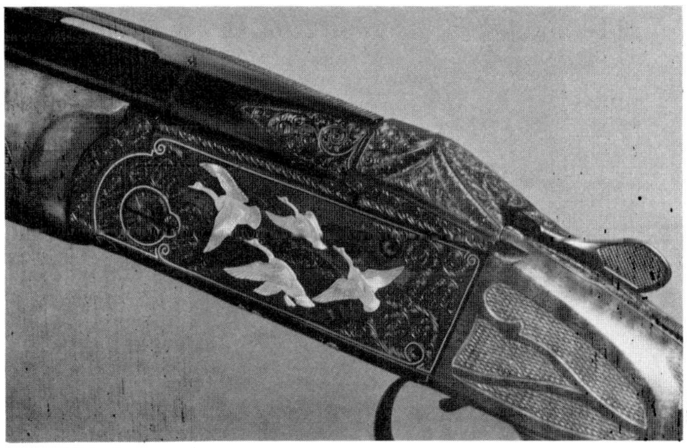

**Wild geese is the subject of Mr. Kolouch's skill as seen here on this view of a Krieghoff Model 32 Vandalia trap gun.** (Photo by Markow, courtesy of Southwest Shooters Supply)

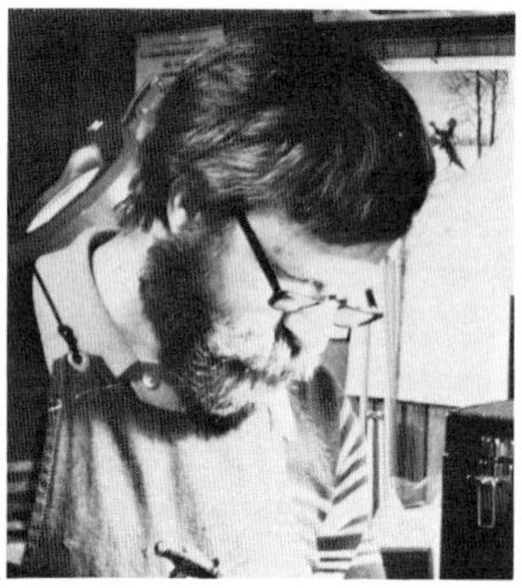

# John M. Kudlas

*Rochester, Minnesota*

Mr. Kudlas was born in Chicago but grew up on a small farm in central Wisconsin. He became interested in engraving as a boy, when he saw an engraved revolver on the cover of a magazine. When he expressed his interest, his father, a Czechoslovakian metalsmith, made him some cutting tools. He practiced engraving as a hobby throughout his school days.

Mr. Kudlas has an extensive formal education, which began with a Bachelor of Science degree in physical education and biology from La Crosse State College. He earned his Master of Science degree at the University of Wisconsin in 1967. His education continued at the University of Utah, where he received his PhD from the College of Health and Leisure Studies. While attending college he worked in welding and machine shops.

Mr. Kudlas has engraved professionally for the past ten years. He is a member of the National Muzzle-Loading Rifle Association and a life member of the National Rifle Association. He spends his leisure time fly-fishing, mountain climbing, or attending gun shows and "mountain men rendezvous."

Using only the chasing hammer and chisel, Mr. Kudlas has engraved over 200 firearms, plus countless inscriptions and decorations on muzzle-loading arms. For example, at the Fort Bridger Rendezvous in 1976 he inscribed over 100 muzzle-loading arms.

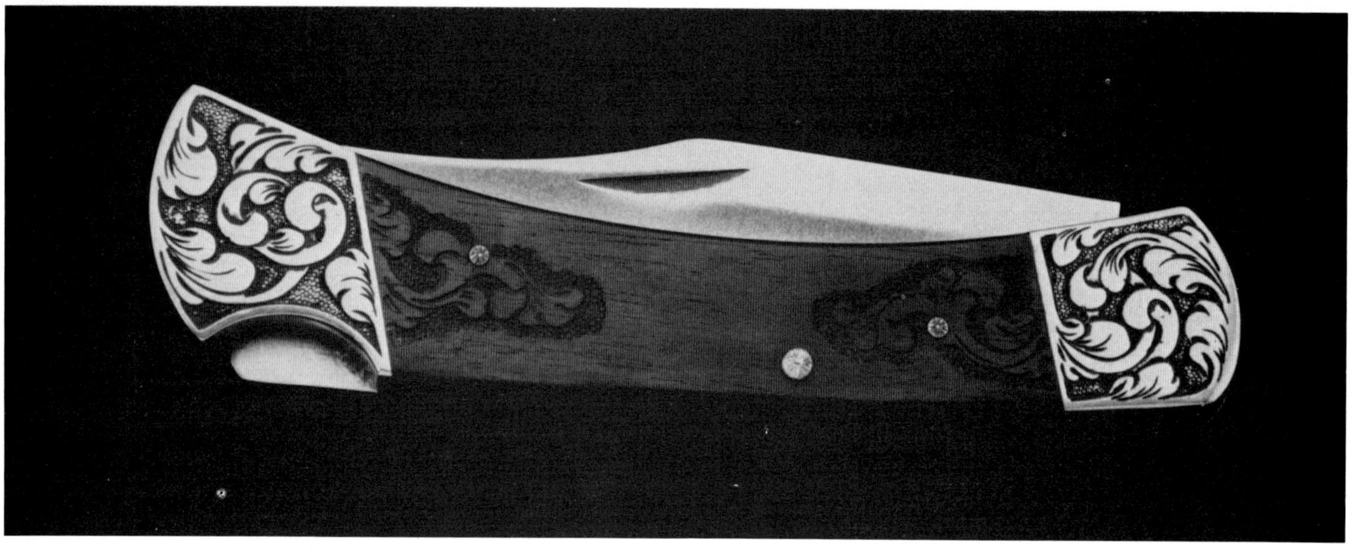

**The bolsters of this Buck folding knife have been engraved with interlocking American scroll, which Mr. Kudlas has continued onto the wood with carving tools.** (Photo by Charles Twing)

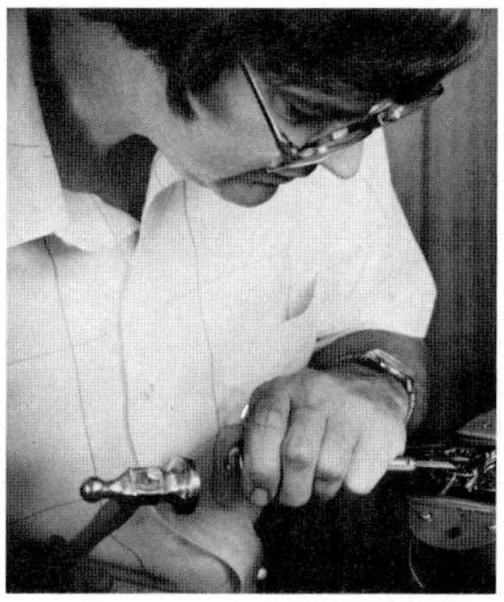

# Ben Lane, Jr.

*Amarillo, Texas*

A native of Amarillo, Texas, Mr. Lane began engraving as a hobby in 1964. His interest in engraving was fostered through another hobby, the collecting of antique arms, as well as his participation in the various shooting sports. In 1966 he began engraving as a part-time vocation and has been a full-time professional since 1968, often putting in as many as sixty hours per week at the block.

Mr. Lane is almost entirely self-taught, yet has achieved exceptional results due in part to his ardent study of both antique and modern engraving techniques. He has authored an article on the subject, entitled "Colt Factory, Single Action Engraving Techniques," which was featured in the March 1978 issue of *Gun Report* magazine.

Mr. Lane has a preference for the engraving of the nineteenth century masters, especially L. D. Nimschke, Cuno Helfricht, and Conrad and John Ulrich. His favorite twentieth century engraver is the late Rudolph Kornbrath, whose works have influenced Mr. Lane's designs.

His other interests include golf, camping, photography, machining, and leathercraft. He is a member of the National Rifle Association, the Texas Gun Collectors Association, and the Amarillo Rifle and Pistol Club, and is a past president of the Pioneer Gun Collectors Association.

Mr. Lane is yet another craftsman who believes that the few arms engravers in America should band together to preserve the knowledge and skills of this ancient art, to prevent its demise despite advancing technology, and to offer alternatives to a population that is being programmed to accept mediocrity.

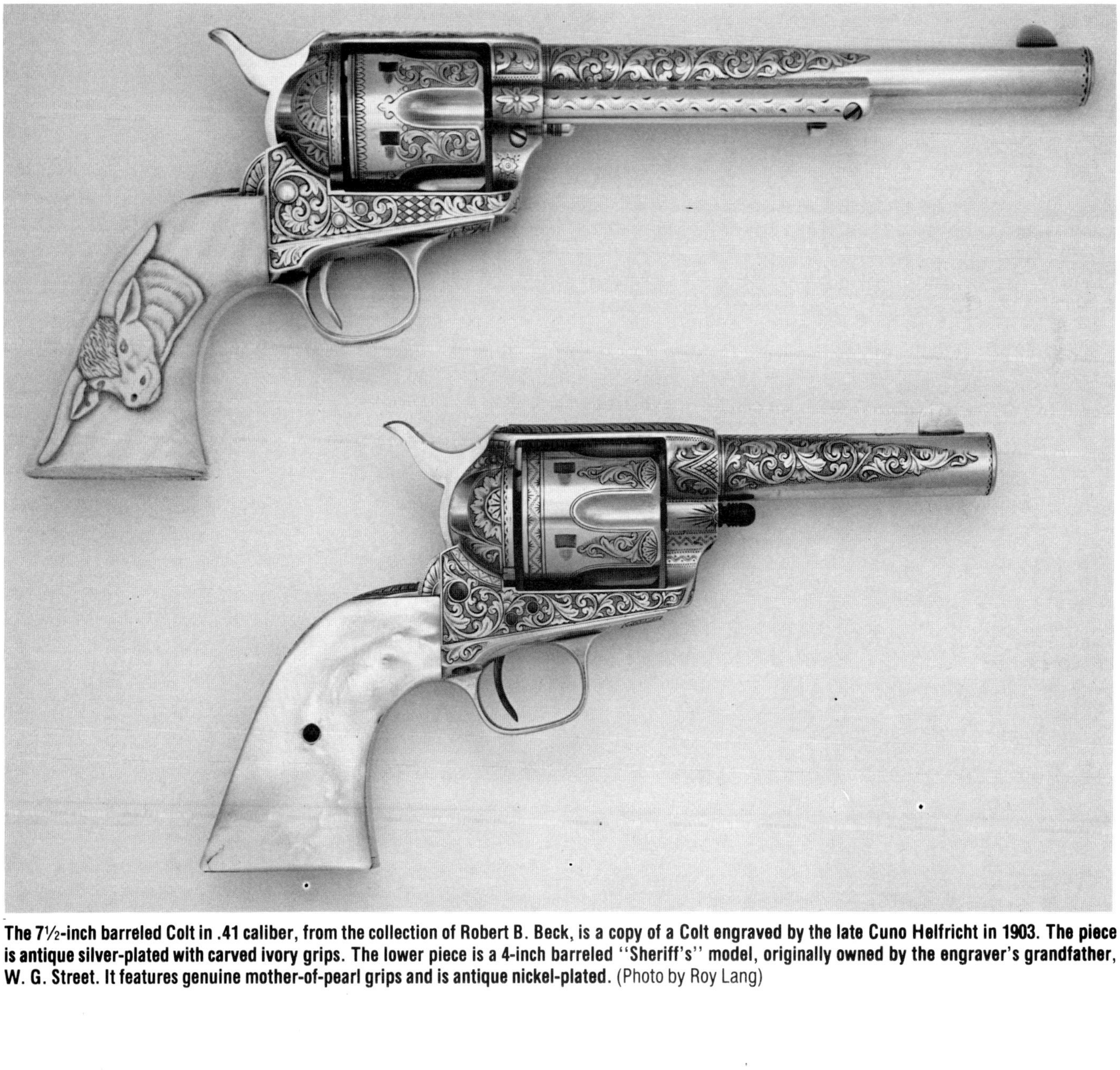

**The 7½-inch barreled Colt in .41 caliber, from the collection of Robert B. Beck, is a copy of a Colt engraved by the late Cuno Helfricht in 1903. The piece is antique silver-plated with carved ivory grips. The lower piece is a 4-inch barreled "Sheriff's" model, originally owned by the engraver's grandfather, W. G. Street. It features genuine mother-of-pearl grips and is antique nickel-plated.** (Photo by Roy Lang)

**A Smith & Wesson Model 27 engraved in semi-relief with a beaded background. Mr. Lane has used Germanic floral scroll plated with antiqued silver.** (Collection of John F. Bickley. Photo by Roy Lang)

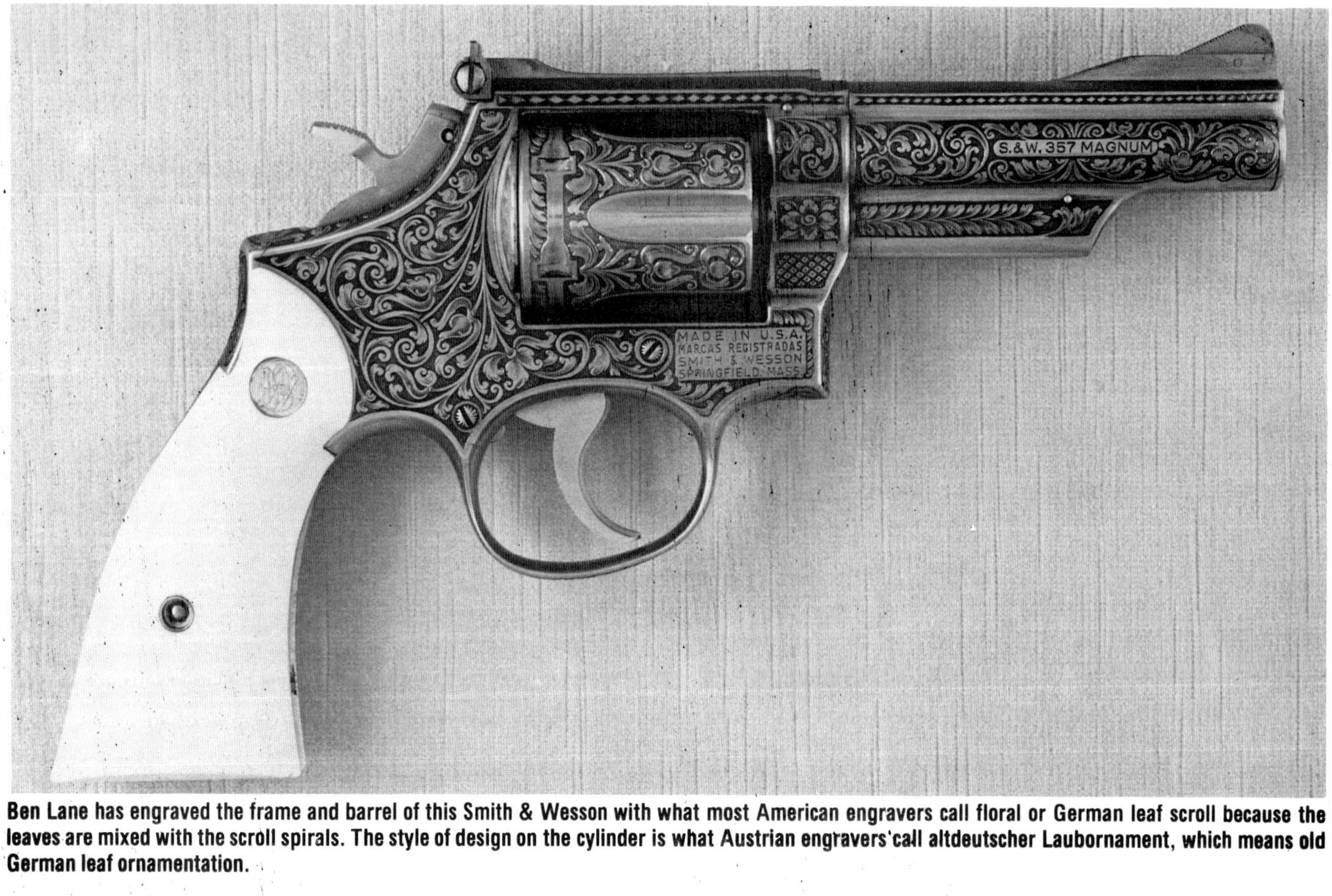


**Ben Lane has engraved the frame and barrel of this Smith & Wesson with what most American engravers call floral or German leaf scroll because the leaves are mixed with the scroll spirals. The style of design on the cylinder is what Austrian engravers call altdeutscher Laubornament, which means old German leaf ornamentation.**

**The receiver of this Model 12 Winchester clearly illustrates Mr. Lane's talent with Germanic floral scroll, done in semi-relief with a beaded background. The Canada goose and pintail duck are inlaid and modeled in silver.** (Collection of Jim Lang. Photo by Roy Lang)

**Mr. Lane used intertwining Germanic floral scroll to offset this humorous scene called "When hunter becomes victim." Note the rifle with broken stock near the bear's paw.** (Collection of Brad Ridgely. Photo by Roy Lang)

**Two sideplates of a Model 1873 Winchester rifle engraved with high relief Germanic floral scroll are shown here. The modeled relief figures have been inlaid in 18 karat gold. The flush borders are a combination of 24 karat gold and silver.** (Collection of Ben Lane. Photo by Roy Lang)

**Germanic scroll in high relief surrounds this African game scene containing a 24 karat gold inlaid elephant in sculpted relief. Flush borders are also of 24 karat gold.** (Collection of Leroy Fortner. Photo by Roy Lang)

**Though this appears to be a lever action rifle, this gun is actually a Model 36 Marlin .410 shotgun which has been decorated by Ben Lane with leaf scroll, flush gold borders and a raised inlaid pheasant.**

**Germanic floral scroll is again the style on this pre-World War II Colt "Ace" .22 caliber pistol. The ivory grips were also monogrammed by Mr. Lane.** (Collection of John F. Bickley. Photo by Roy Lang)

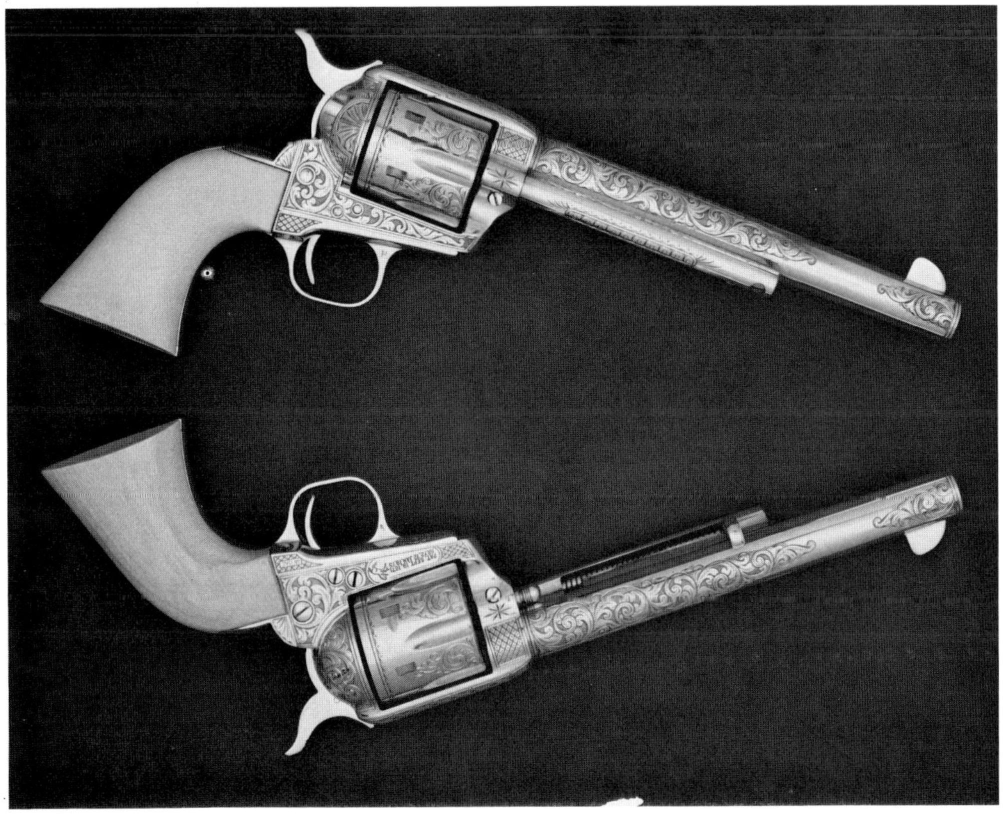

**This brace of Colts has consecutive serial numbers and is bright silver-plated. The design was copied by Mr. Lane from an early Colt engraved by the late Louis D. Nimschke.** (Photo by Roy Lang)

# Edward Joe Machu, Jr.

*San Antonio, Texas*

In his junior year at Alamo Heights High School, in San Antonio, Texas, Mr. Machu (pronounced Ma-who) enrolled in an industrial cooperative training program, in which he attended school half of the day and learned a trade the other half. This was the beginning of his nine-year career as an industrial engraver (die sinker) and machinist.

After high school he joined the Texas Air National Guard, where he learned meteorology as an air weather technician. By the time of his discharge in 1972 he had attained the rank of staff sergeant. During this time he studied art in night school.

Die sinkers work from art designed by another person, and Mr. Machu felt the urge to be more creative. This led him to the shop of fellow Texan Frank E. Hendricks, featured earlier in this text. After practicing for another year to adapt his skills to the work done on firearms, Machu went to work full-time for Mr. Hendricks. Then after three years Mr. Machu went into business for himself.

Among his interests are hunting, fishing, baseball, and bowling. He even considers the forty-five to sixty hours per week spent in his shop a way of having fun. His shop is located within "The Sportsman's Bailiwick," one of San Antonio's fine sporting goods stores.

Virtually all of Mr. Machu's commissions are from private individuals. He estimates he has engraved approximately 400 guns during his career. He works entirely with hammer and chisel except for background relief and the modeling of figures, where he uses dental burrs in conjunction with a flexible shaft motor.

Along with engraving, Eddy (as he prefers to be called) Machu is also skilled at making pistol grips of wood and ivory, as well as at carving and scrimshaw. His work can be seen at the many arms collector shows he attends throughout the state of Texas.

Bas-relief American scroll surrounds a high relief game scene on this Browning Superposed.

The receiver of this 12 gauge Browning Superposed is plated with frost nickel and engraved by Mr. Machu with American scroll surrounding high relief steel scenes.

Seen here is a small gauge Iver Johnson Skeet-er shotgun that has been brought to Mr. Machu for engraving. This piece features Nimschke style American scroll, finished in antique silver with blued barrel.

The receiver of this 12 gauge Browning Superposed is plated with frost nickel and engraved by Mr. Machu with American scroll surrounding high relief steel scenes.

Enlarged detail of this Browning receiver shows what flush inlaid figures look like to the engraver before he cuts in the detail.

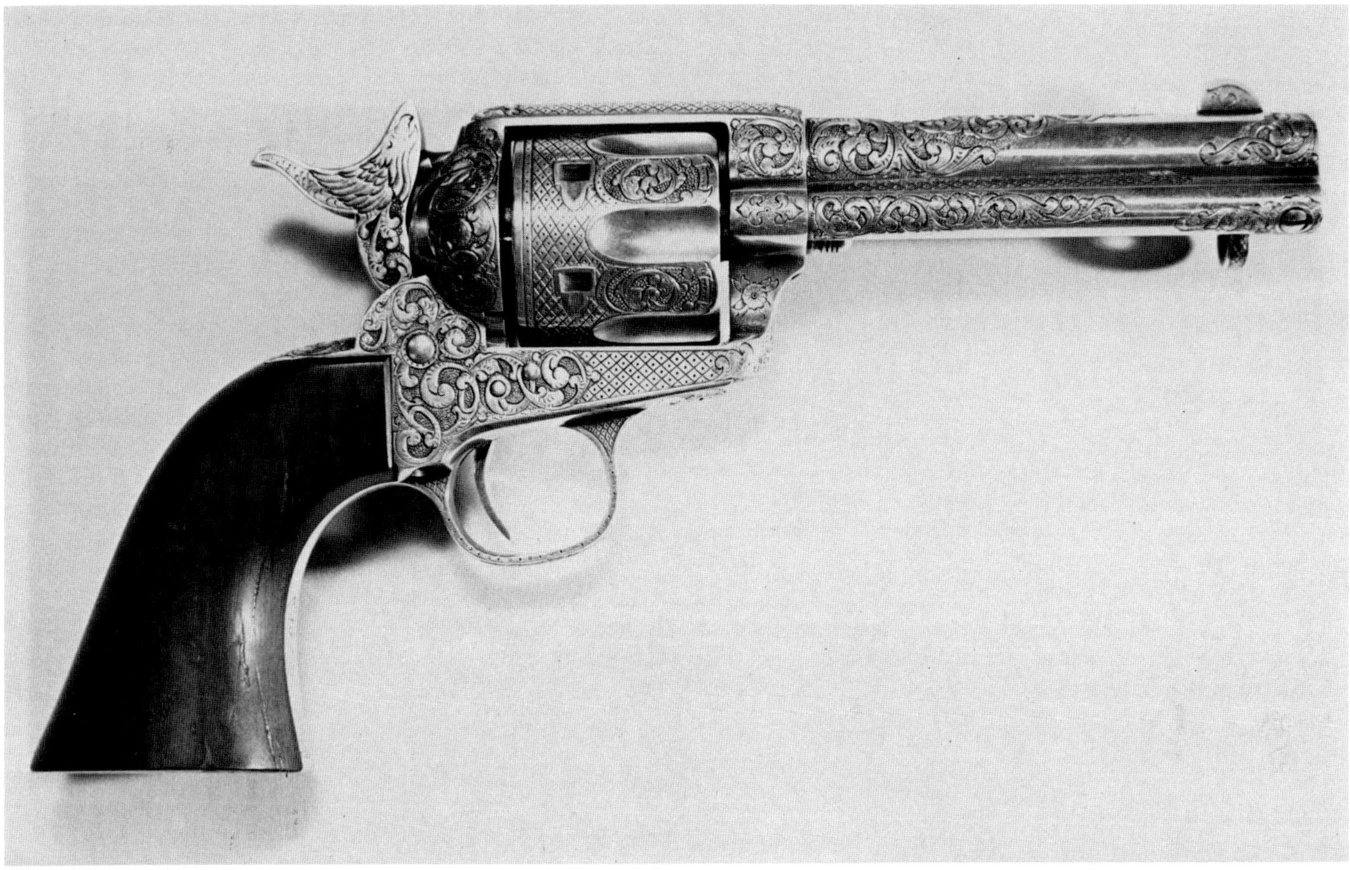

Engraved for a close friend in 1973, this early Colt "Peacemaker" features Nimschke style scroll, extensive crosshatching, and an eagle head hammer. The one-piece ebony grips are also by Mr. Machu.

The influence of the late Louis D. Nimschke can be seen in the scrollwork engraved into the receiver, lever and barrel of this Ruger single shot by Ed Machu.

# William H. (Bill) Mains

*Las Vegas, Nevada*

A person of many skills, Bill Mains served an apprenticeship as a goldsmith from 1952 to 1953; apprenticed as a watchmaker from 1953 to 1954; and then studied lathe work and jewelry engraving at the Bowman Technical School. He worked as a locksmith from 1959 to 1962, and as a machinist in 1963. Mr. Mains saw military service from 1955 to 1958 and attained the rank of airman first class with the United States Air Force.

He is a contract engraver for several arms makers, and his work is often featured in books and magazines, where it is depicted as factory engraving. Over the years he has done engraving for Charter Arms, Colt, High Standard, Ithaca, Mossburg, Sterling, Thompson Center, and Winslow.

After his training as a jewelry engraver in 1954, Mr. Mains developed a rotating footwheel engraving vise that resembles the ancient potter's wheel. This foot control eliminates the necessity of moving the vise with the hand while working with hammer and chisel and allows the continuous cutting of a scroll from beginning to end.

In addition to arms work Mr. Mains engraves dies and stamps, jewelry, and musical instruments, and also does wood carving. He is interested in the study of wildlife and is a member of the National and International Wildlife associations, the Defenders of Wildlife, the National Rifle Association, and the Texas Gun Collectors Association.

During his career as arms engraver, Mr. Mains has done commissions for many personalities, including General Suharto, Liberace, Elvis Presley, and Sammy Davis, Jr. He estimates he has engraved over 1,600 firearms during his career.

**Winchester Model 21 displaying nonrelief Austrian scroll, deep relief scenes, and gold birds.** (Courtesy of Wally Hicks)

**Ljutic Bi-matic with seven gold animals and deep relief Germanic scroll and scenery.** (Courtesy of Bill Barrenger)

**An early work of Mr. Mains, done in 1967. He has cut relief German scroll and game scenes into this blued Model 12 Winchester.**

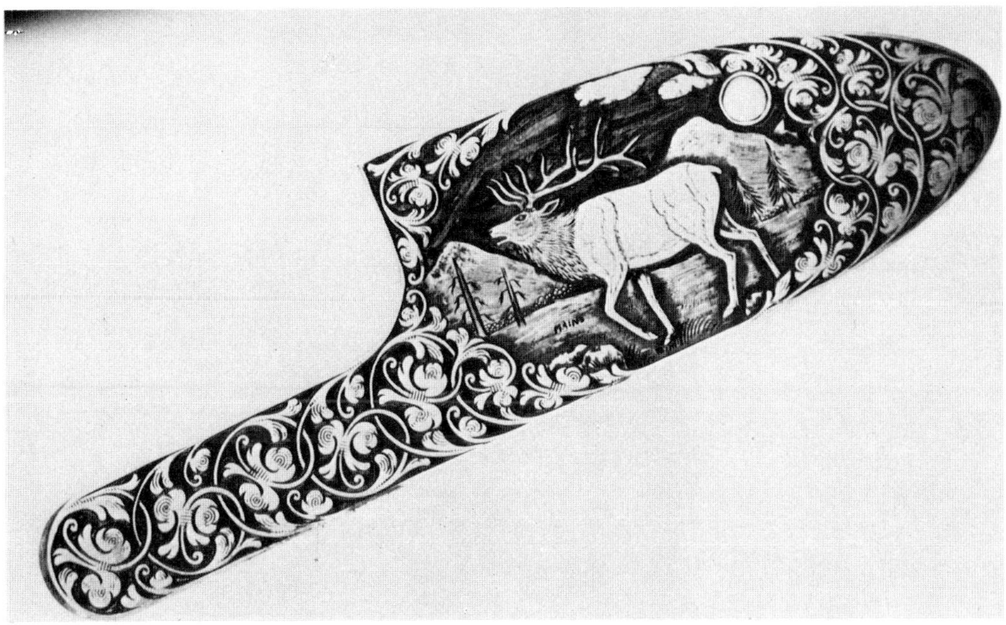



**This sideplate from a double rifle was recently engraved by Mr. Mains with relief Germanic floral scroll. Note the signature: MAINS, to the left of the elk's front leg.**

Winslow Emperor grade rifle which appeared on the cover of the 1968 edition of *Guns Annual*. The rifle displays eight gold animals and relief gold oak leaves. Only two of these examples were made.

Remington Model 1100 with five gold animals, deep relief scenery and low relief scroll.

Rudy Marek has been engraving since the 1940s, when he made his first graver from a saw blade. He later received some instructions from a Czech jeweler who engraved trophies, but like most hand engravers he did not know how to engrave steel. When he traveled to Chicago in the 1950s to meet the late Arnold Griebel, he found the master arms engraver lacking in cooperation. But at least he was able to see the master's tools, which was of some help.

Mr. Marek received his first paid commission in 1956 and has been engraving part-time ever since, while working full-time as a police officer. In the late 1960s Mr. Marek worked two years for the Colt factory. At present, he engraves ten to twenty hours per week. In addition to firearms, he also engraves cremation urns.

As a hobby Mr. Marek sews period costumes. He is a life member of the National Rifle Association and a member of the Oregon Arms Collectors Association.

## Rudy W. Marek

*Banks, Oregon*

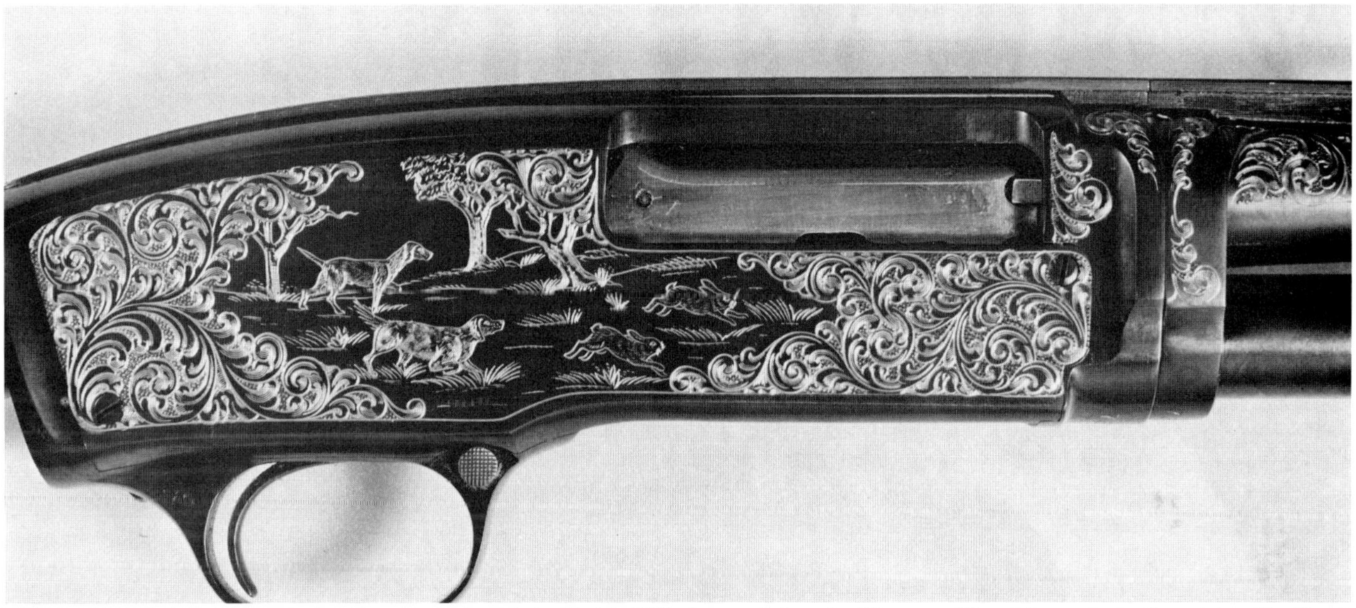

Mr. Marek's work is shown here on the receiver and barrel of a Model 42 Winchester shotgun in .410 gauge. Though he works in other styles, Mr. Marek has chosen a typically American style of scroll decoration with beaded background for this piece. This example illustrates the appearance of a gun engraved through the original bluing.

A combination of German style scroll and leaf designs decorates the receiver of this Ljutic trap gun, engraved by Rudy Marek in 1967. Note the trap target incorporated in the design.

# Lynton S. M. McKenzie

*Tucson, Arizona*

Lynton McKenzie was born at Rockhampton, Queensland, Australia in 1940 and became a United States citizen in 1973. He was educated at King's College in Sidney, the oldest school in Australia.

At the age of thirteen he took up muzzle-loading shooting and gunsmithing. His main interest was the restoration of fine antique British sporting arms. He eventually served as a technical advisor and instructor for the various Australian muzzle-loading clubs and represented his country as a contestant in the international matches against England and South Africa.

During this time he worked as a photoengraving etcher while attending night courses to learn all the other allied printing trades. He not only became the leading photoengraver and arms restorer in Australia, but also taught himself to engrave arms with burin, hammer, and chisel. Since he was the only arms engraver in the country, and had no one to teach him more, he left for Europe. He carried letters of introduction from leading Australians, including the Prime Minister. These gained him access to major arms works and museums, where he was able to make detailed examinations of select items.

Mr. McKenzie eventually settled in London. He states that he planned to stay six months, but wound up staying six years. During this period of his life he worked seventy hours per week engraving for such famous gunmakers as Holland & Holland, Rigby, Westley Richards, John Wilkes, Cogswell & Harrison, and Chaplins, but most of his work was done for James Purdey and Sons. At Purdey he worked with Ken Hunt, whom he considers to be the finest arms engraver that England has ever produced. During their close association, Hunt and McKenzie shared many ideas and techniques that eventually helped to improve the overall look of British arms engraving.

As an arms restorer Mr. McKenzie has received work from all over the world. Of royal arms alone he has restored antique pieces belonging to the French from Louis XIII to Louis Napoleon, as well as arms from crowned heads of England, Spain, Portugal, Holland, Austria, Germany, Italy, and Russia.

Many persons in the United States complain of declining quality and lack of skilled work, yet make no move to correct such trends. Stanley Diefenthal of the New Orleans Arms Co., however, not only encouraged Mr. McKenzie to work in the United States, but also provided every possible facility to enable him to do the best quality of work. Thus, between the years of 1970 and 1978 the New Orleans Arms Co. was able to produce a remarkable number of museum quality cased arms with beautiful accessories that equalled or excelled anything produced in Europe. Most of these projects were accomplished on arms of American origin.

In 1978 Mr. McKenzie moved to Boulder, Colorado, where he closely collaborated with master gunsmith Richard Hodgson. In 1980 he moved to Tucson, Arizona.

**A bighorn sheep from a painting by Bob Kuhn is engraved on the side of this Ruger No. 1 rifle with decor by Mr. McKenzie.**

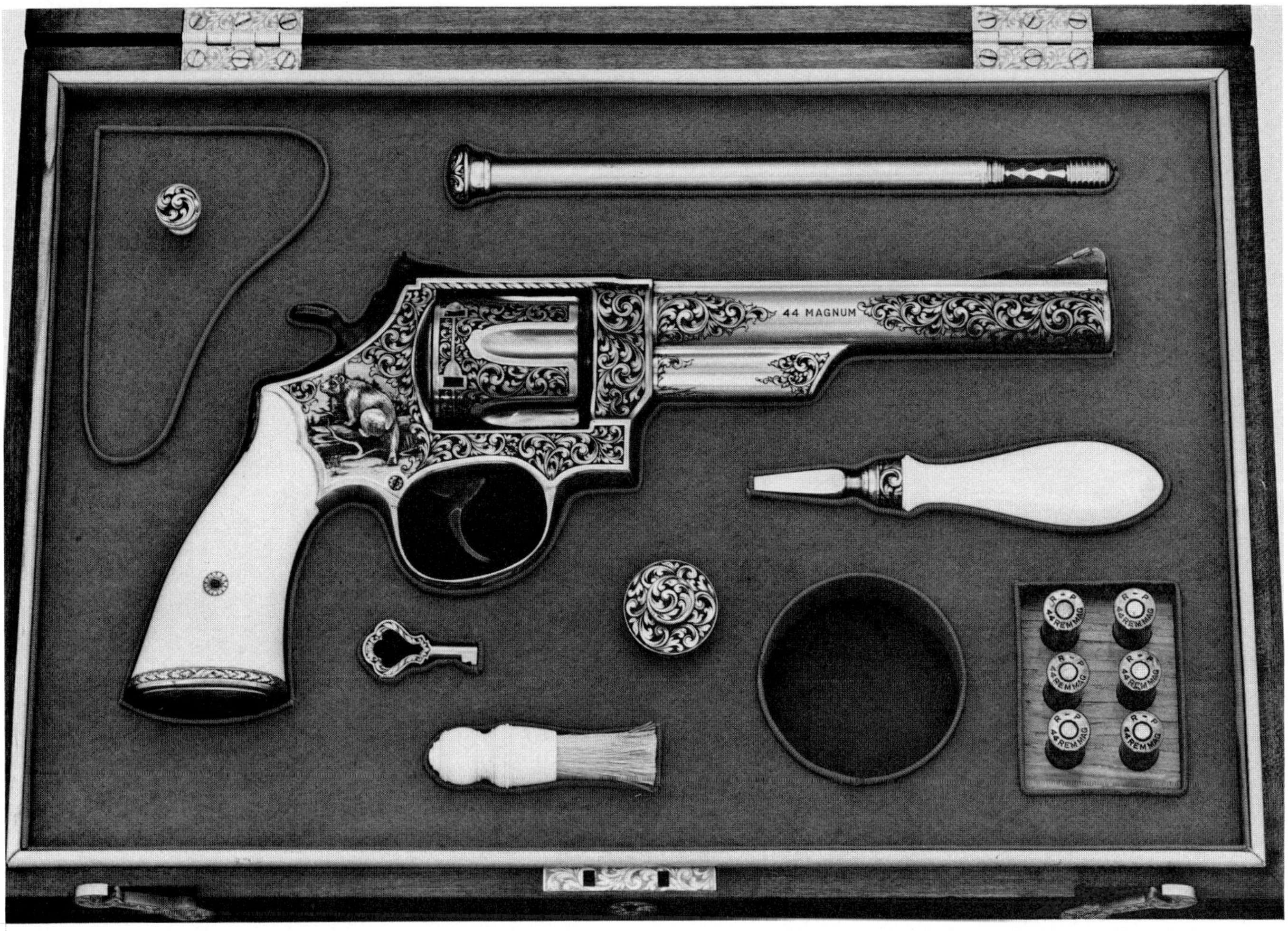


A richly decorated Smith & Wesson Model 29 is seen here with case and accessories, as well as decoration by Lynton McKenzie. Note the ivory patch container is missing in this photograph.

A Colt Government model pistol made for President Ferdinand Marcos of the Philippines. The slide of the gun exhibits the flag and shield of the Philippines in raised gold; a beautifully executed intertwining line motif in flush gold appears near the muzzle. The grips, case, and all metalsmithing were done by Richard Hodgson.

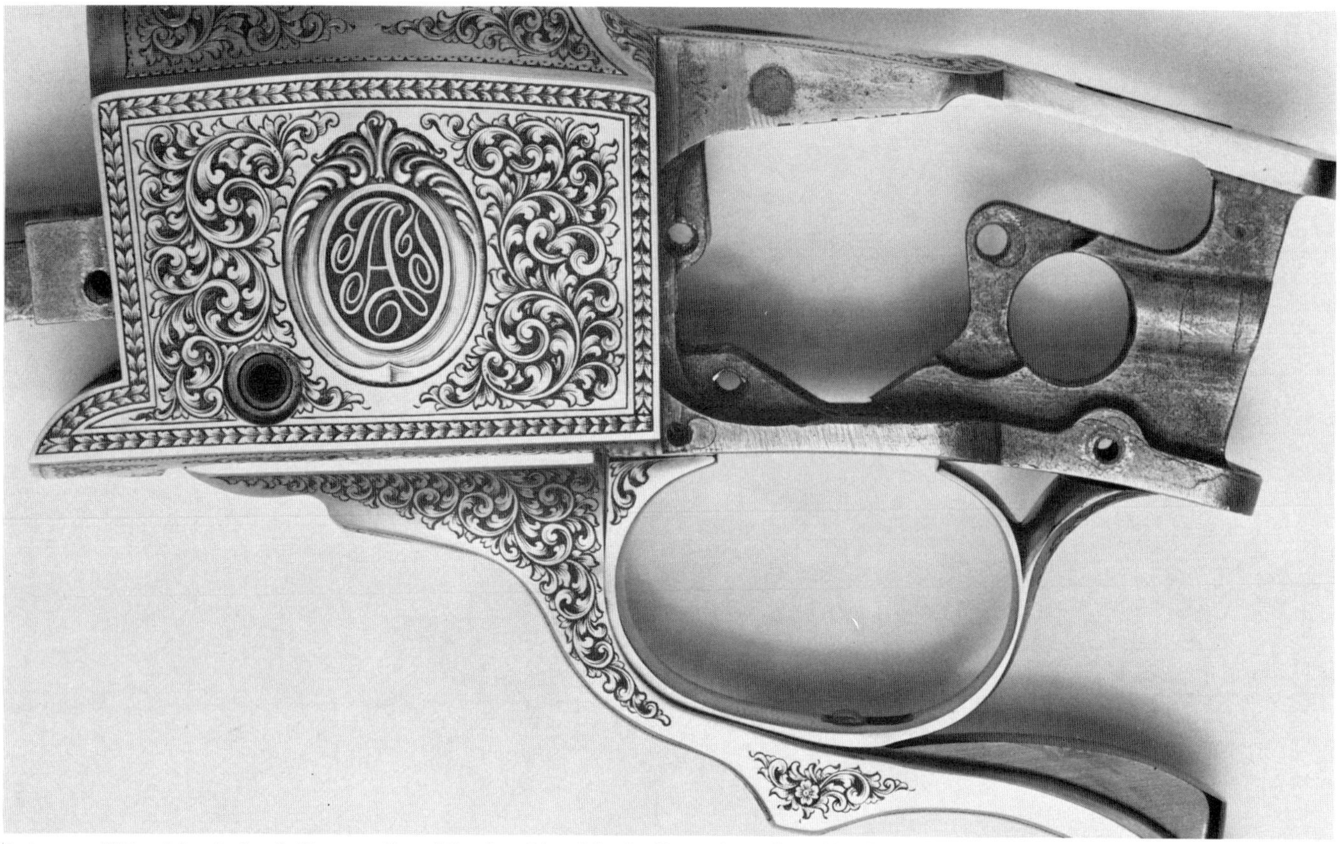

**Made as a gift for John Amber (editor emeritus of the** Gun Digest**) by the Ruger Arms Co., this piece was engraved by Lynton McKenzie with an oval inlay on each side by Winston G. Churchill. This piece was featured in the 1976 edition of the** Gun Digest.

**Originally commissioned by Dr. Simon J. Conrad with the assistance of publisher Wallace Beinfeld, the Colt shown here in detail enlargement is one of a set of five Colts on the subject of the American West. The entire set is featured in R. L. Wilson's** Book of Colt Engraving. **The theme of this design is "The Settlers." The gun is shown here before bluing.**

This .44 magnum Smith & Wesson by Lynton McKenzie features game scenes inspired by the paintings of Bob Kuhn.

Detail of the Bowie knife, Colt derringer, and the oil bottle and patch container of the Colt-Winchester set. Note that the wheat chaff or running leaf border around the finger guard of the knife is a solid gold inlay.

Another presentation piece for President Marcos of the Philippines is this Colt single action, featuring raised gold shield motif and running leaf pattern on the loading gate. The finish is French gray.

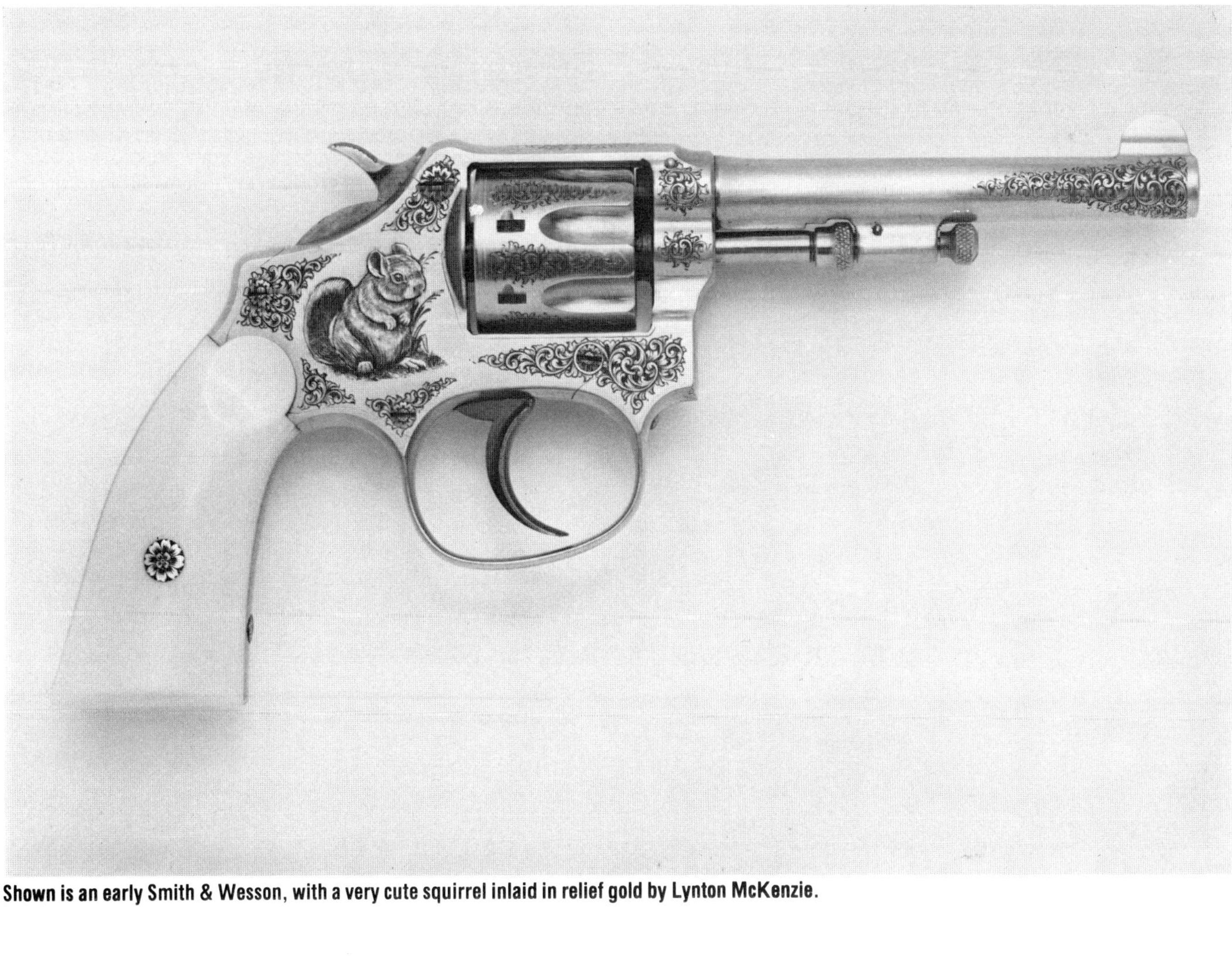

Shown is an early Smith & Wesson, with a very cute squirrel inlaid in relief gold by Lynton McKenzie.

Underside of an English Purdey shotgun shown in detail. Although several other engravers have begun to imitate this style of scrollwork, the author refers to this type of work as "McKenzie scroll," because Mr. McKenzie introduced this style in the United States. Note how the trigger plate screw has been incorporated into the design.

Detail enlargement of the left sideplate of an 1873 Winchester engraved by Mr. McKenzie. The photo shows the fine detail of the raised relief gold eagle motif. Note the special fixture made by Mr. McKenzie to hold the sideplate in his vise. He has seven complex fixtures for the Colt single action alone.

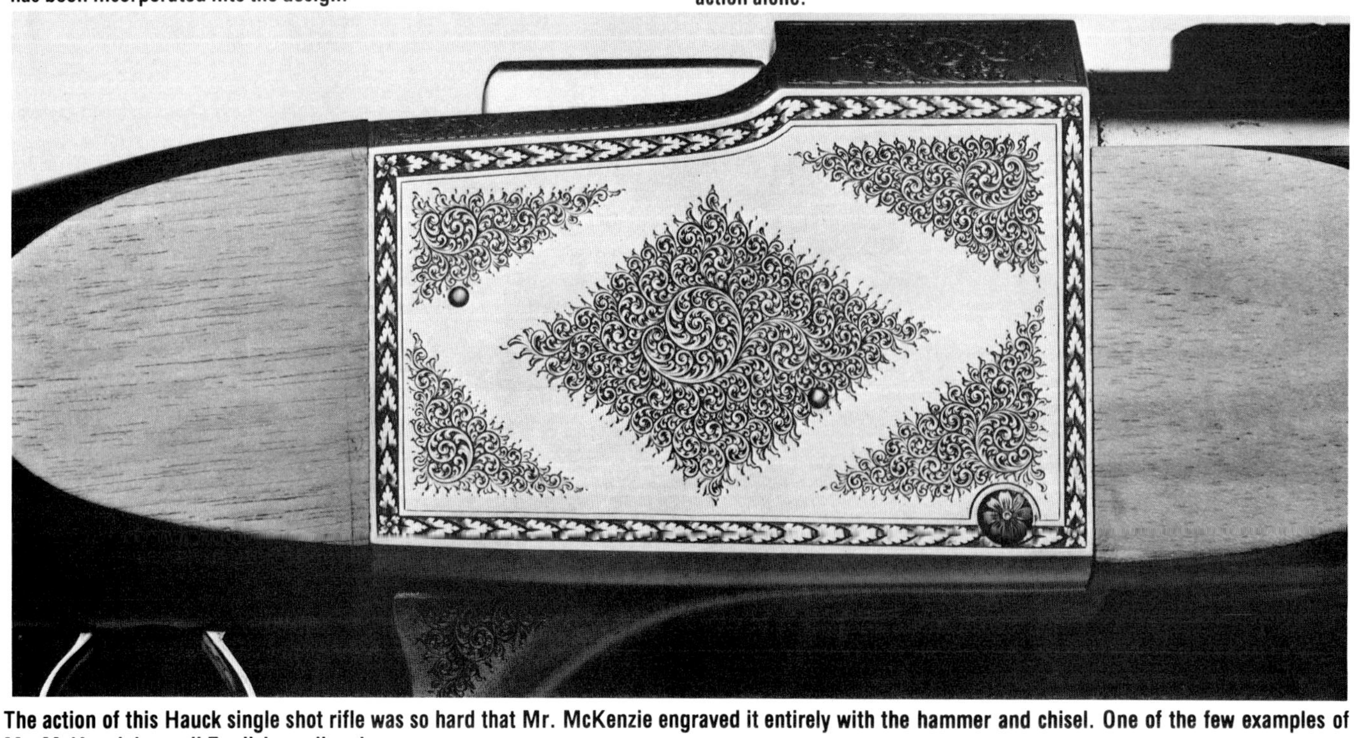

The action of this Hauck single shot rifle was so hard that Mr. McKenzie engraved it entirely with the hammer and chisel. One of the few examples of Mr. McKenzie's small English scrollwork.

A trio of superbly crafted folding knives by Horn, embellished in Lynton McKenzie's characteristic relief scrollwork.

# Konrad Oker, Sr.

*Crystal Lake, Illinois*

Though originally trained as a machinist, Konrad Oker has been near the engraving trade for all of his forty-two years. His father, Herman Oker, Sr., has been an engraver for sixty years and was trained in Schwabisch-Gmund, Germany. Though young Konrad was encouraged to learn engraving at an early age, it was not until he was older that he developed the patience that engraving requires.

Mr. Oker, who works full-time as a plant superintendent, is able to devote about thirty hours per week engraving and still have a little time left over to engage in his favorite pastime of big game hunting.

Since the middle 1970s Mr. Oker has studied diligently with his father to improve his skills and feels that his engraving has come a long way in that time. Though his father is retired from his die engraving business, he still does a few guns per year, working only for a single customer. At present Konrad uses both the traditional hammer and chisel and the Gravermeister in his work.

Konrad Oker is a Past Master of Masonic Lodge 169, Crystal Lake, Illinois.

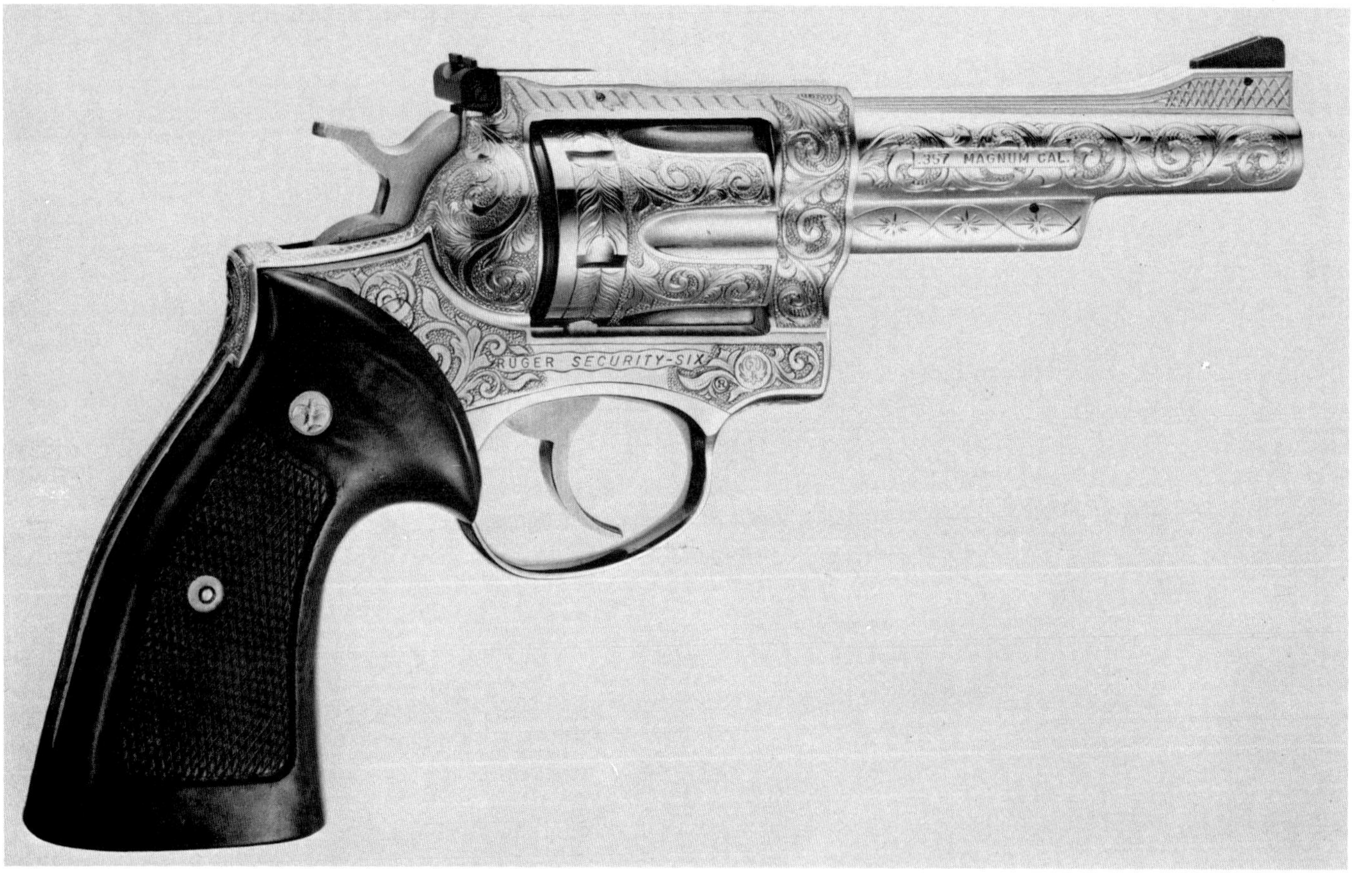


**A background beaded with a circle punch is used to contrast the scrollwork on this Ruger stainless steel revolver.** (Collection of Konrad Oker)

Mr. Oker tells us that he inlaid fifty-four inches of gold wire into the borders of this Llama automatic engraved with scroll and leaf designs. The inlaid bear's head is modeled of 18 karat gold.

# Marcello Pedini

*Dix Hills, New York*

Born in 1940, Mr. Pedini is a native of Orvieto, Italy. He attended technical and engraving school in Milan, where he also served an apprenticeship of eighteen months with B. Falloni & Brothers.

A few years later Mr. Pedini began his own free-lance engraving business and in 1972 moved with his family to Australia. His work became known quickly and was popular in shooting circles in Melbourne and other cities. Eventually the Pedini family became Australian citizens.

As his work became more widely known, he was invited to relocate in the United States. He settled in Worcester, Massachusetts, where that city's museum of fine arts recognized his work as that of a "master engraver."

Mr. Pedini is skilled in the use of the burin, the hammer and chisel, and all other traditional tools. During the past twenty-five years he has employed every technique of hand engraving, including sculptured relief, inlays, bank note scenes and figures, as well as dry point and etching.

In January 1979 he was asked to join Regency House Art, Inc., of Plainview, New York. He is now the artist responsible for the engraving division of that company, where he works in conjunction with wildlife artist and painter Guy Coheleach.

Among Mr. Pedini's works are the engraving of the dies for the bronze bells of the Concilium Oecumenicum Vaticanum Secundum, in Rome, Italy and the coining dies for the Montreal Olympic Games coins.

**The head of this gold-inlaid duck has been magnified 124 times to illustrate Mr. Pedini's attention to detail on inlaid figures. This fine example is on a Holland & Holland shotgun.** (Courtesy of Regency House Art)

Mrs. Pierce began learning the skills of arms engraving in 1976 through the help of fellow Oregonian, Bob Valade (featured later in this text). Like Mr. Valade, Mrs. Pierce has worked since 1978 for Gerber Legendary Blades on their special presentation series of knives. During her career as engraver she has worked on rifles, handguns, and shotguns as well as knives. She estimates she has engraved 150 firearms altogether.

Mrs. Pierce is interested in motorcycling, hunting, and deep-sea fishing. She is also a member of the National Rifle Association.

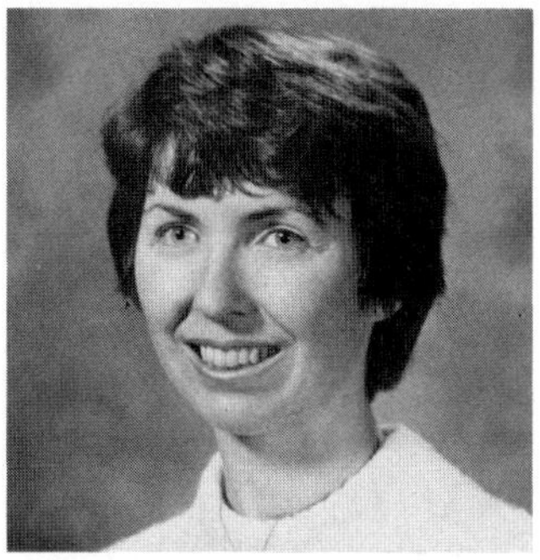

# Barbara R. Pierce

*Hermiston, Oregon*

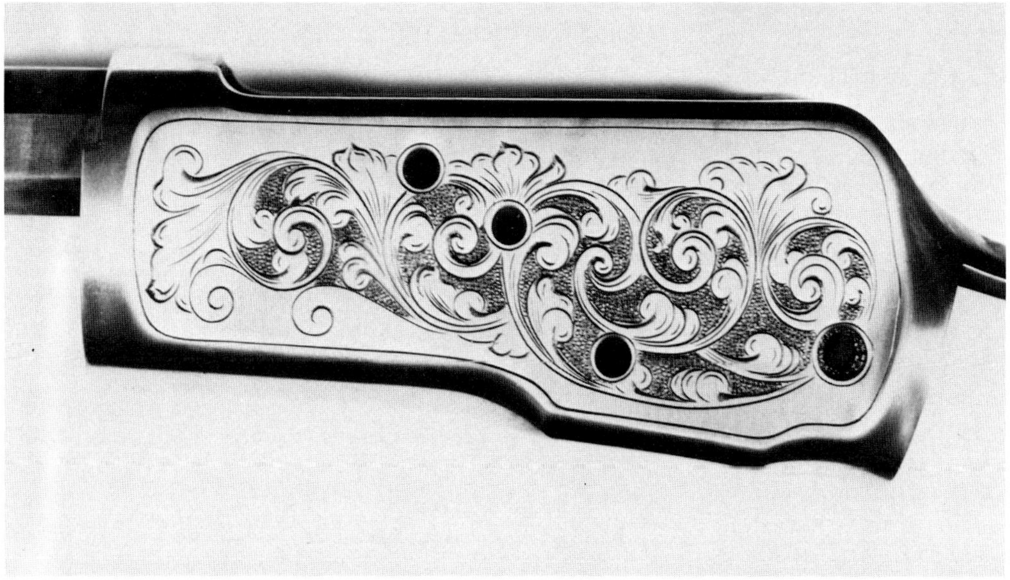

**Mrs. Pierce has engraved the receiver of this Winchester Model 1892 rifle featuring an American scroll in bas-relief and a background beaded with a circle punch.** (Photo by Lyons Photo Studio)

# Paul R. Piquette

*Chicopee, Massachusetts*

Paul Piquette began his career as an engraver in 1976 at the Smith & Wesson factory after submitting a portfolio of his work to factory historian Roy Jinks. His apprenticeship was served under the watchful eyes of long-time Smith & Wesson engraver Russ Smith (featured later in this book), who introduced him to the basics of the trade. Soon afterward he developed and began working in his own special style.

Mr. Piquette works exclusively with the chasing hammer and chisel, either while standing at a bench vise or sitting at the engraver's block. In addition to Smith & Wesson's work, he also engraves long guns, knives, and jewelry at his home shop.

Among his most ambitious works is a profusely inlaid Smith & Wesson .44 magnum revolver that he recently engraved in cooperation with Russ Smith and Domingos Jocquin. This gun recently sold at auction for $80,000. The proceeds of this sale were used by Smith & Wesson to help fund the United States Olympic team.

Among this young artist's interests are painting, photography, fishing, model building, music, and Bible study. He is a member of the Springfield League of Artists and the National Rifle Association.

**These three enlarged views of a Smith & Wesson revolver show Mr. Piquette's bas-relief American style scrollwork. The scroll has been backgrounded by cutting with a parallel line graver, instead of the more common method of backgrounding with a punch. The right sideplate features a bald eagle inlaid in gold. The head of a wolf has been inlaid into the crane, with a black bear inlaid on the left side of the frame.**

# Wayne E. Potts

*Denver, Colorado*

Like Winston G. Churchill, featured earlier, Wayne Potts began expressing himself as a youngster by carving his own toys from wood; and drawing was one of his favorite pastimes.

Early in life Mr. Potts was trained as a dispensing optician, a trade that he followed for forty-seven years until his retirement in 1978. He had a thirty-five month change of pace during World War II, when he participated in five major Pacific campaigns as a sergeant in the United States Marine Corps.

Mr. Potts is mainly self-taught as an engraver. Originally he cut with the hammer and chisel. Then he met John Rohner, who introduced him to the Gravermeister, which he now uses in preference to the older methods. Since his retirement, he engraves about thirty hours per week and decorates knives and jewelry, as well as firearms.

During the summer months, Mr. Potts spends most of his time at his mountain cabin near Creede, Colorado, where he enjoys hiking with his dog, fishing, photography, and of course engraving.

**Bas-relief scrollwork is combined here with floating oak leaves to balance the design. Note the "Nimschke" stars at the tops of the frame and barrel. On this piece Mr. Potts has filled the engraving and stamping with gold paint.** (Collection of Ron Julian)

**Mr. Potts has engraved this stainless steel Smith & Wesson with what one might call a floating oakleaf design in relief. The grips are stag.** (Collection of John Andrews)

## John Vest

**Ruger No. 1 engraved by John Vest in 1978. The inlays are of white and yellow gold which accent the animal's natural markings.** (Collection of Lloyd Jensen. Photo by Bill Havlina)

**Freedom Colt set A number 2, dedicated by engraver Dwain Wright to Walt Whitman. The gun is gold-plated, fully engraved and features a steel portrait of Whitman on the left recoil shield, relief chiseled in 7/32-inch depth.** (Collection of Dwain Wright)

# Dwain R. Wright

**Set A number 3 of 20 from the Freedom Colt series is dedicated to Abraham Lincoln. Coverage is about 80 percent of the metal surfaces and includes Presidential seals signed by former President Ford. The revolver displays semi-relief scrollwork; a full relief portrait of Lincoln appears on the recoil shield.** (Collection of Amado Hernandez)

# Marcello Pedini

This floor plate from a Champlin rifle was engraved to exactly duplicate a painting by wildlife artist Guy Coheleach. Executed under twenty-five power magnification, the inlaid ram and rocks are done in a combination of gold and silver. Mr. Pedini's unusual style of floral decoration surrounds the game scene. The accompanying photo of Mr. Coheleach's painting illustrates the connection between the two works. (Photo courtesy of Regency House Art and Guy Coheleach)

# Dwain R. Wright

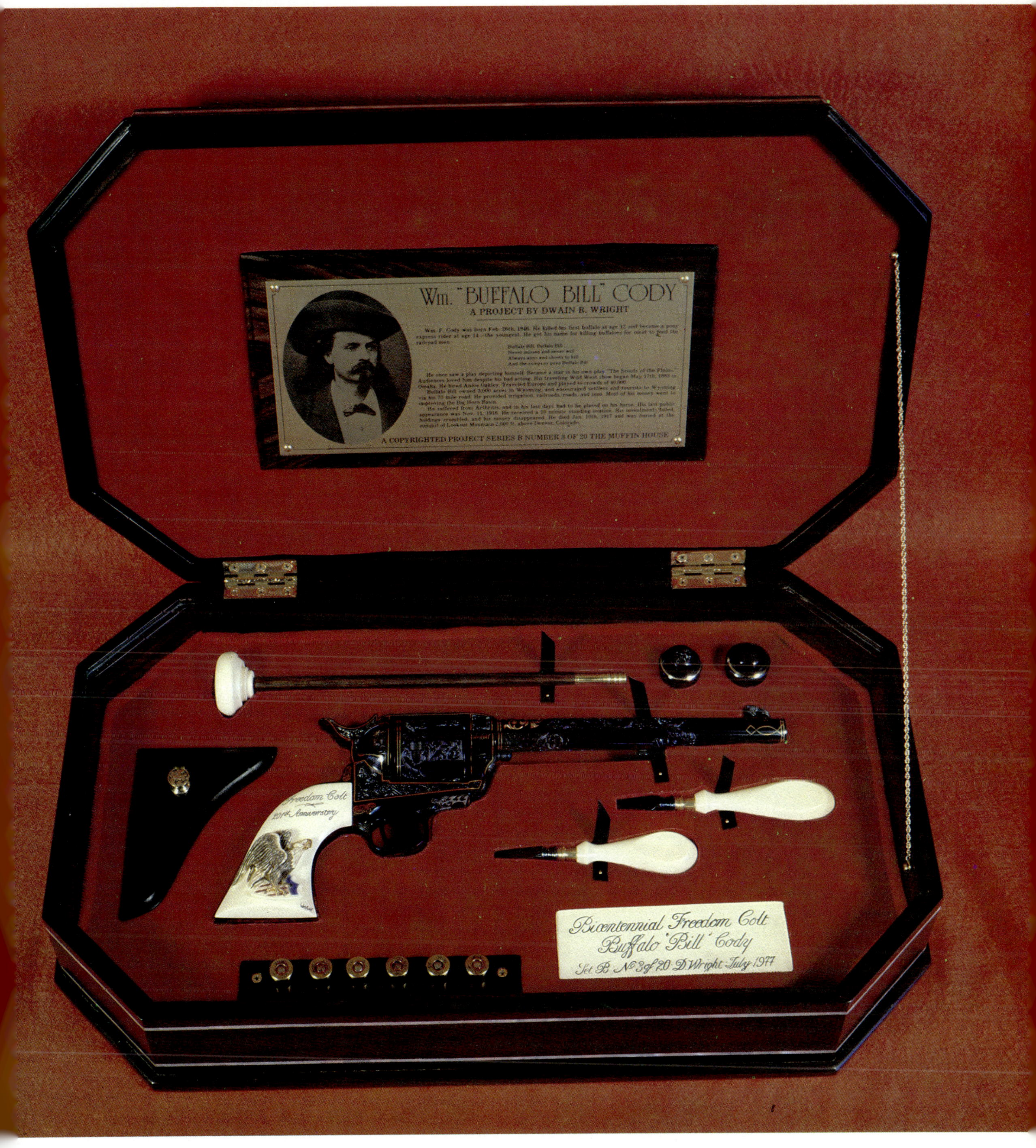


**Set B number 3 of 20 from the Freedom Colt series is dedicated to Buffalo Bill Cody. This gun has inlays of 24 karat gold. The Buffalo Bill inlay weighs one ounce. The oil bottles are gold-inlaid and engraved. Approximately 490 hours were required to complete this involved outfit.** (Collection of Fred Fiet)

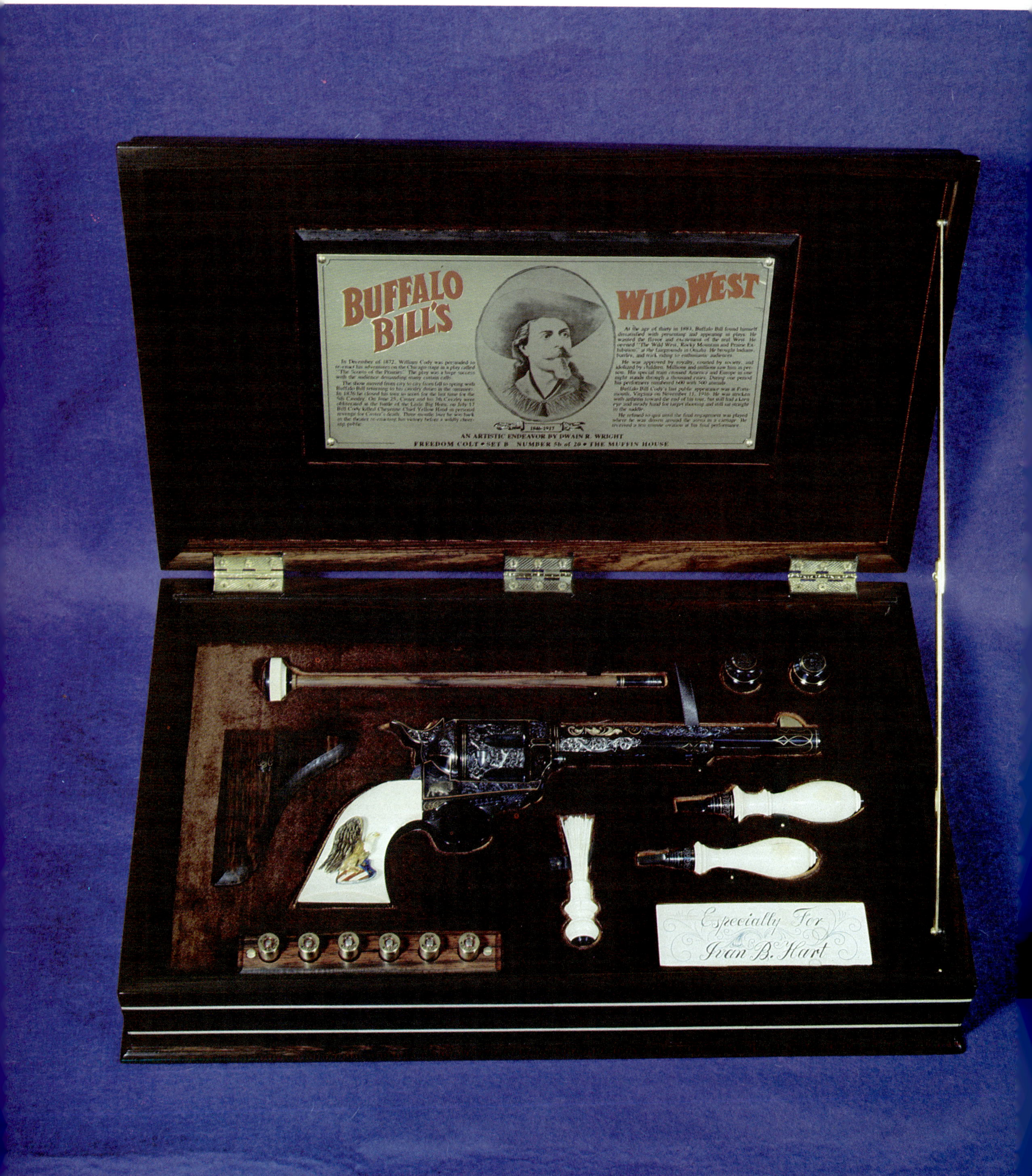


**From Dwain Wright's Freedom Colt series, set number 5 of group B. This deluxe outfit was dedicated to Buffalo Bill Cody. Note the combination of gold and scroll decoration, plus the colorful and unusual ivory grips.** (Collection of Ivan B. Hart)

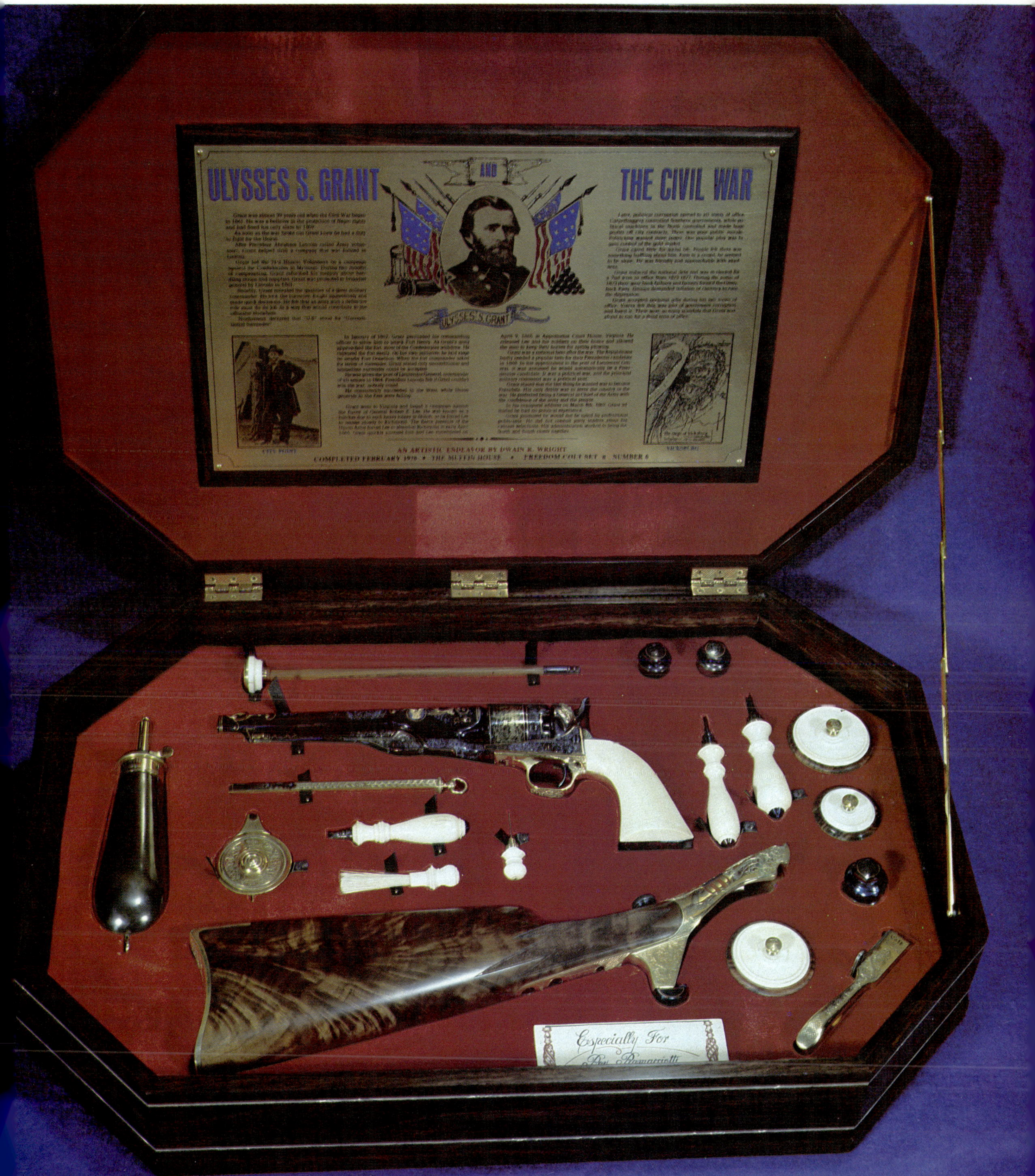


**One of the more elaborate of the Freedom Colt arms, set B number 6 is an 1860 Army Colt which has been engraved, gold inlaid, charcoal blued and gold plated.** (Collection of Ray Ramaciotti)

Set A number 3 of the Freedom Colt series features a relief bust of Abraham Lincoln. The grips are inlaid with the presidential seal and color-scrimshawed by Mr. Wright. (Collection of Amado Hernandez)

Jack Prudhomme is well known in at least three roles. To gun collectors he is best known as an engraver of long standing, who has plied his trade in his native Shreveport since the 1940s.

He is also an author, having compiled and written the first and second editions of *The Gun Engraving Review*. His books, long out of print, are highly sought after by book and arms collectors today.

His third area of renown is as a professional wrestler from 1932 until 1948. In 1938 he held the light heavyweight title under the ring name of Jack Hagen.

During World War II, Mr. Prudhomme served as a master sergeant in the Army Air Corps. He is noted for having designed and built the first two-seat P51 Mustang, an aircraft that carried General Eisenhower over the front at Normandy.

Mr. Prudhomme states that his recognition as a wrestler helped him to meet many fine engravers of the 1930s. These men assisted him in learning more about the art and included R. J. Kornbrath, Arnold Griebel, and Max Bruehl, all now deceased.

An unusual honor and distinction came to Mr. Prudhomme in 1973 when the R. W. Norton Art Gallery exhibited fifty-six firearms he had engraved and inlaid. The arms (all on loan from their owners) represented twenty-seven years of Mr. Prudhomme's career.

Mr. Prudhomme's other interests include photography, golf, hunting, and bowling. He is a member of all Masonic bodies and a Shriner, as well as Past Master of his lodge.

## E. C. (Jack) Prudhomme

*Shreveport, Louisiana*

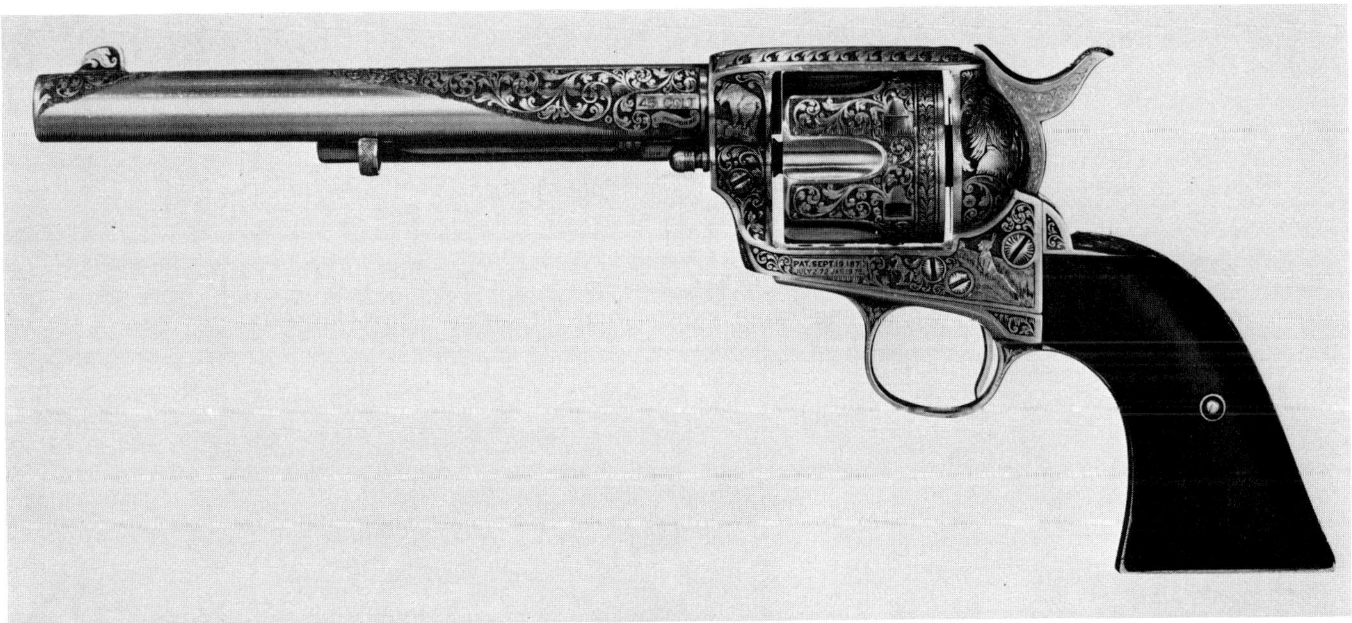

A Colt 7½-inch scroll engraved single action with scenes of animals and Indians.

This rifle floor plate exemplifies the complete coverage offered by E. C. Prudhomme. The Bull elk and background scenes are relief-engraved.

A Colt Second Model dragoon, profusely engraved and inlaid with two colors of gold. The scenes of American Indians and scouts are inlaid in yellow gold and the bands on the cylinder are yellow gold with unusual green gold inlay of fleur de lis.

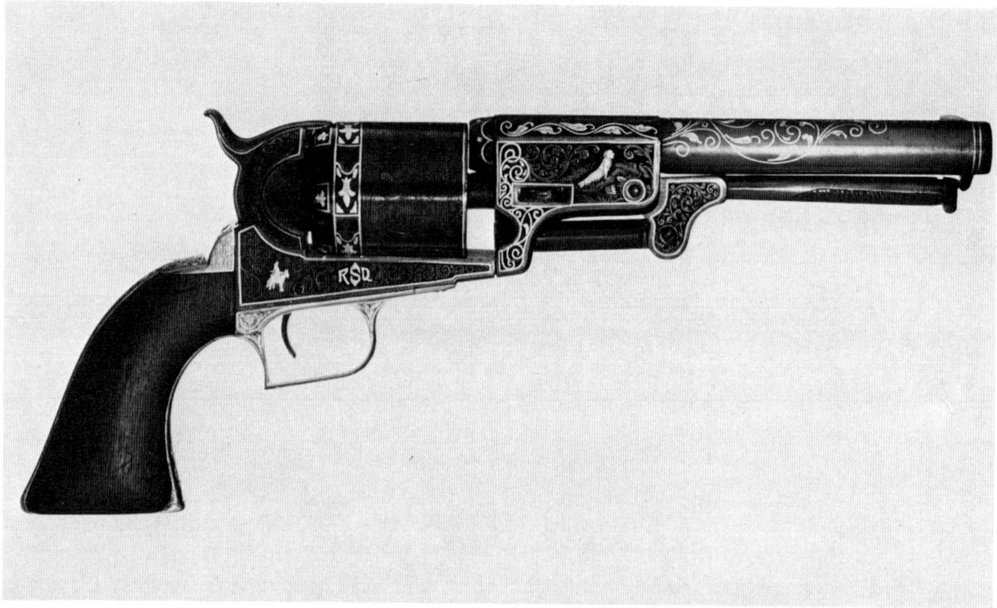

A Samuel Colt 150th Anniversary Commemorative standard model with full Prudhomme coverage. The considerable amount of gold, in three different colors, is inlaid in a fluid style of floral coverage.

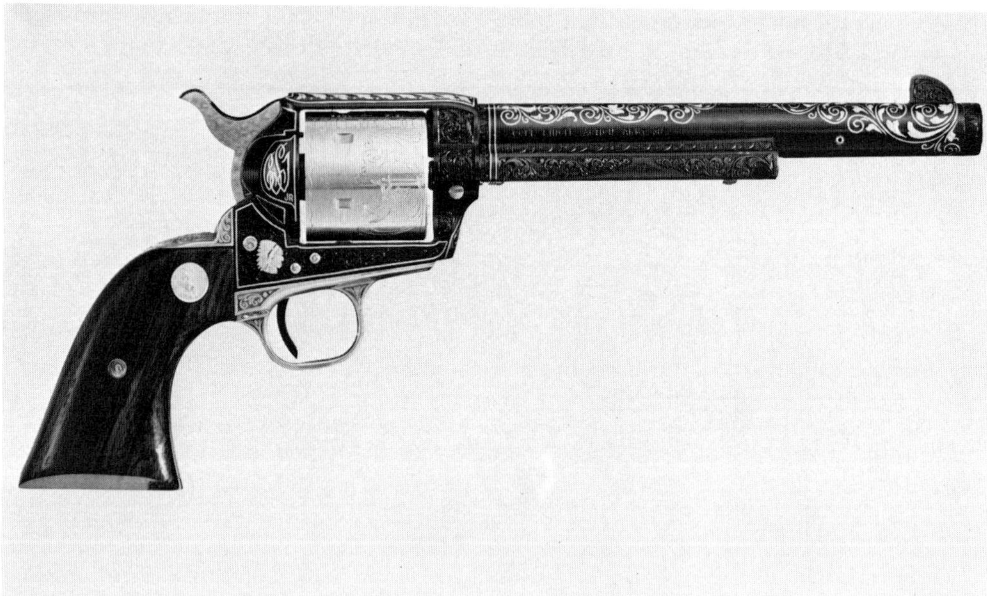

Left side of a Remington Model 32 shotgun. The scroll engraving contrasts with the gold-inlaid scene of oak leaves and ducks landing.

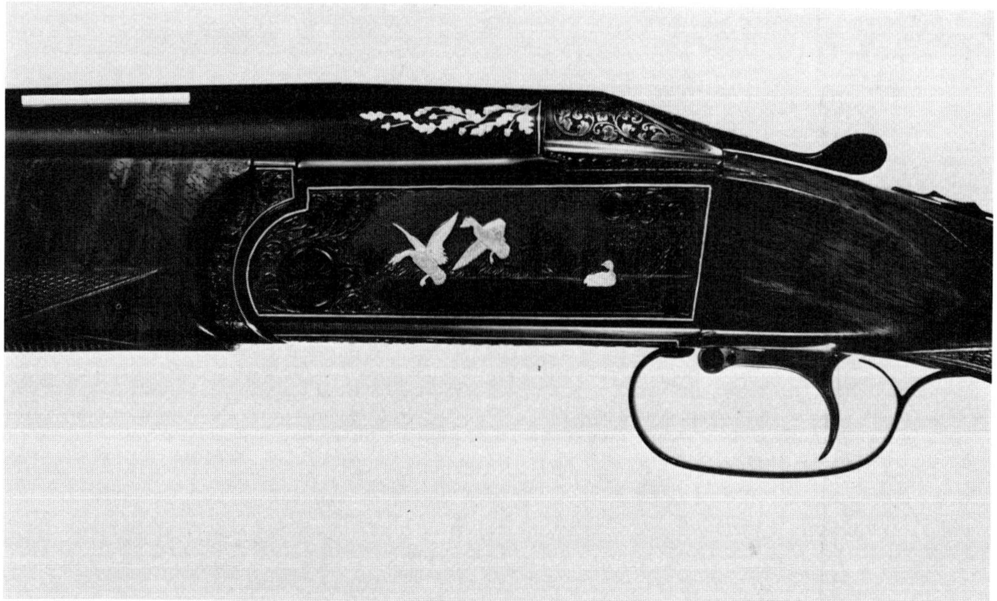

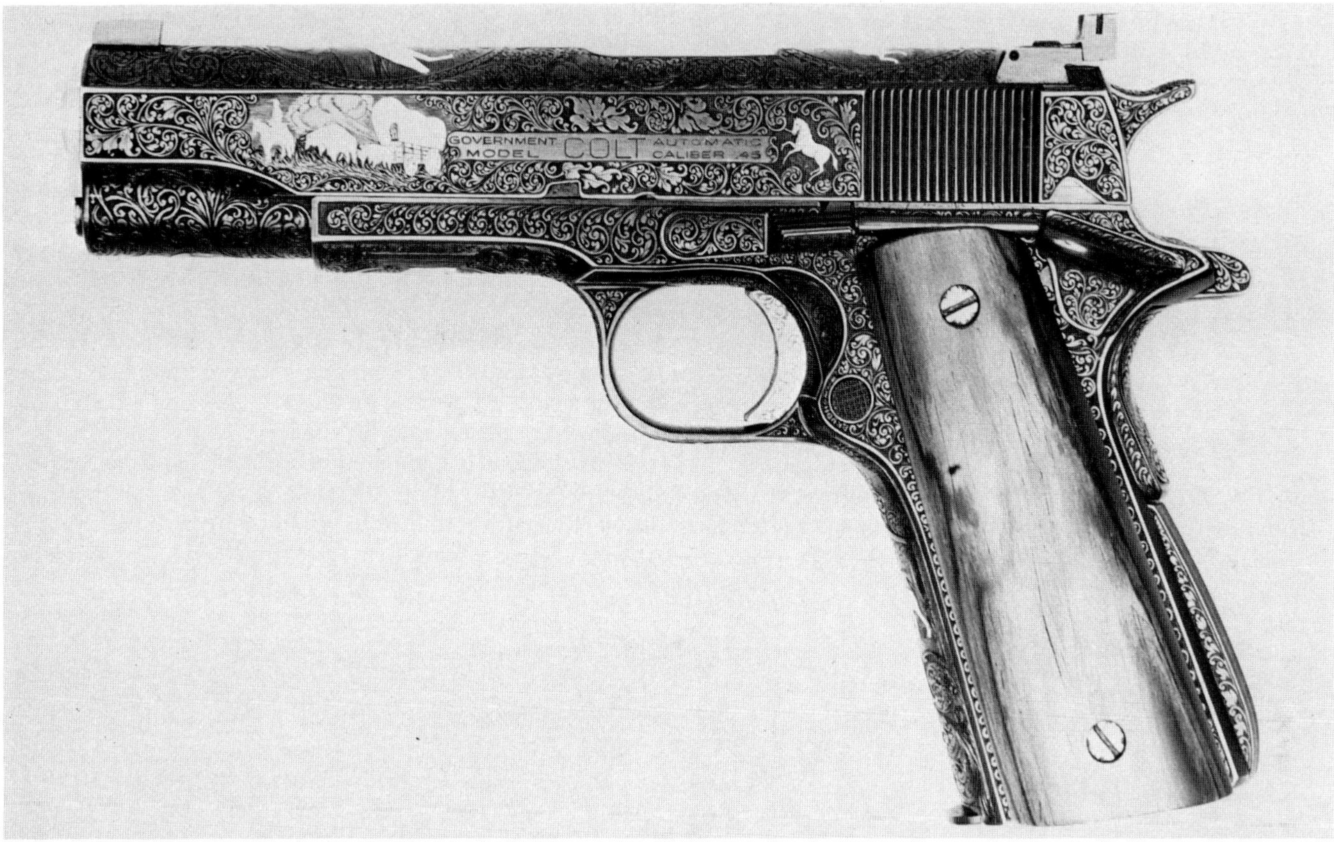


This piece, executed in 1956, has been profusely decorated with medium relief Germanic scroll incorporating some floral motifs. Difficult to see in this photograph are the extensive flush border inlays, done in both gold and platinum. Inlaid on this pistol are eight gold and platinum figures, including animals, cowboys, and an ox-drawn covered wagon.

Made in 1907, this Colt 1905 automatic was engraved and inlaid by Jack Prudhomme in 1963. The mustang and rampant colt on the left side are modeled in relief from the base metal, but the borders and a monogram, RWN, are inlaid in three colors of gold. Note the signature on the slide just above the trigger: ENGRAVING BY E.C. PRUDHOMME 1963.

# Martin Rabeno

*Ellenville, New York*

Born and raised in New York City, Mr. Rabeno received a thorough art education. As a teenage student he attended the High School of Music and Art and was able to spend much time visiting the arms and armor section of the Metropolitan Museum of Art. Upon graduation he received the J. Henry Holloway Medal from the International Graphic Arts Teachers Association.

Afterward he attended college where he earned a Bachelor of Fine Arts degree with a major in painting. After graduation he became a high school art teacher in the Catskill Mountains. Mr. Rabeno continued his education at the State University of New York College at New Paltz, where he received his master's degree in gold- and silversmithing.

Always interested in firearms, Mr. Rabeno began engraving guns in 1975, beginning with muzzle-loading rifles that he made. At present he engraves on a part-time basis, working very closely with his clients.

Mr. Rabeno is a member of the National Rifle Association and the National Muzzle-Loading Rifle Association. He also participates in the activities of the British and Hessian Armies of the Revolution, in the 64th Regiment of Artillery.

**Interlocking relief scrollwork has been used on this Model 27 Smith & Wesson in conjunction with a howling wolf scene.** (Collection of Ed Jacobs)

**Mr. Rabeno has engraved this flintlock with simple line engraving, similar to that found on many period pieces.**

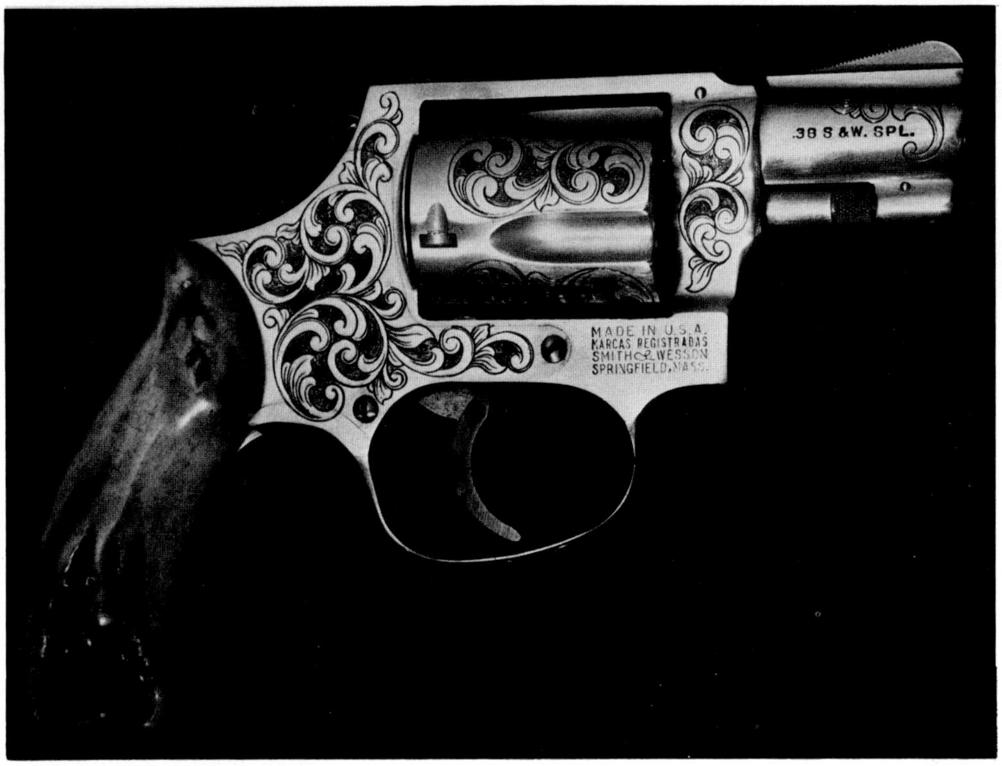


**Bold, borderless scrollwork accents the surfaces of this Model 60 Smith & Wesson made of stainless steel. Mr. Rabeno has matted the background with a single point punch.** (Collection of Leo Cecil)

**The slide of this Colt Government model illustrates Martin Rabeno's talent at designing and executing interlocking American scroll in high relief. Note the shell motif ahead of the ejector port and the engraver's initials below the rear sight.**

# John & Hans Rohner

*Boulder, Colorado*

Since 1962, John Rohner has been the Curator of Museology at the University of Colorado Museum. He holds a Bachelor of Arts degree from the University of Iowa, where he majored in zoology. His education was interrupted by World War II, from 1943 until 1945. He served in the Pacific theatre as a corporal with the 4th Marine Division and was wounded on the island of Saipan. Then he continued his education at the University of Iowa, and in 1950 received his master's degree in museum methods.

As an engraver John is basically self-taught, but states that his early efforts were assisted by James B. Meek, the author of *The Art of Engraving*, the only American text on how to engrave firearms. For cutting tools he uses both the hammer and chisel and the Gravermeister, a pneumatic chisel that operates like a miniature jackhammer. Mr. Rohner and his brother-in-law Don Glaser are codevelopers of this device, which is now used by several engravers. Mr. Rohner has also assisted many beginners who wanted to learn arms engraving. Among these people is his son Hans, who now works with him.

Hans Rohner began to learn engraving at the age of fourteen and has been practicing this skill for seven years. Hans, like his father, engraves part-time; each spends about twenty hours per week at this activity. Hans is employed full-time as an instrument maker at the Joint Institute for Laboratory Astrophysics.

Both Rohners have been regular exhibitors at the annual National Rifle Association Convention, where they demonstrate engraving with the Gravermeister to the thousands who attend yearly.

In addition to engraving, John Rohner also engages in wildlife photography, gold- and silversmithing, and bronze sculpture, and collects North American Indian artifacts and artwork. Somehow with all his other interests, he still finds time to be an official measurer for the Boone and Crockett Club.

Hans Rohner is following in his father's footsteps when it comes to engraving. He spends his spare time as an amateur telescope and toy maker.

**Featuring stock work by Leonard Brownell and a barrel by the late H. M. Pope, this .22 caliber Remington "Rolling Block" target pistol has been adorned with Germanic scrollwork by John Rohner.**

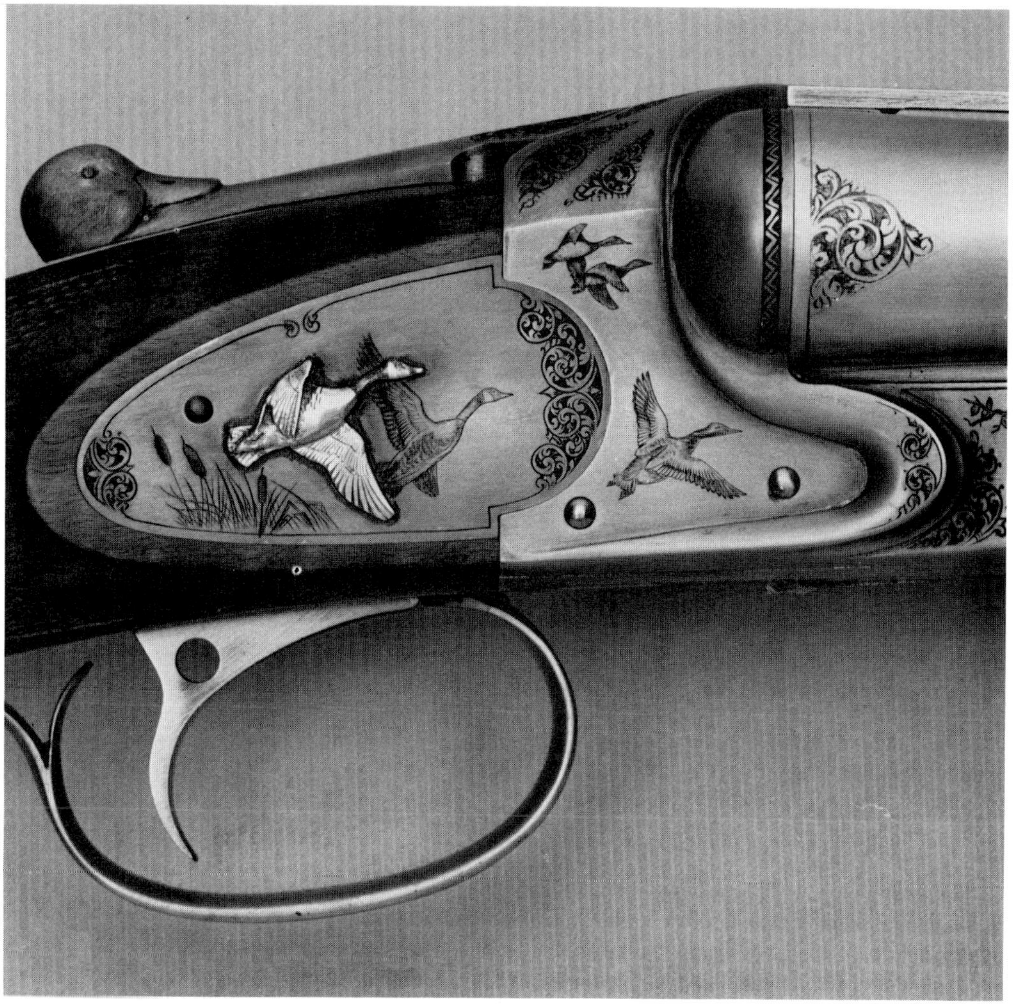

**Shown in two views is a Winchester Model 21 custom-made by Leonard Brownell of Wyarno, Wyoming. Special features by Mr. Brownell include the stocks, sideplates, and metal sculpturing, including the duck's-head opening lever. The engraving is shown still in progress and features German scroll and modeled relief inlays by John Rohner, with fine line wildlife and scenery by son Hans. The original bird drawings were done by Maynard Reece especially for this gun.** (Collection of Maynard Reece)

# Alvin E. Scott

*Baytown, Texas*

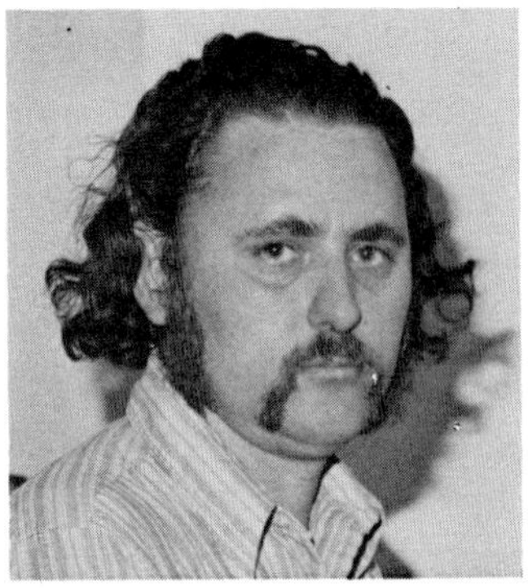

One of several arms engravers who is native to the "Lone Star State," Mr. Scott left school in the eighth grade to pursue employment in a variety of jobs. His hope was to someday make a living by working with firearms, but his chance didn't come until 1974 when he was employed as a gunsmith by master gunsmith Wayne Barnett. A year later Mr. Scott met master engraver Frank E. Hendricks, who was a great help in developing Scott's skills as an engraver.

Mr. Scott later moved to Alaska, where his engraving business flourished until the pipeline boom ended. He returned to Texas in early 1979, where he continues to engrave at the Pasadena Gun Center.

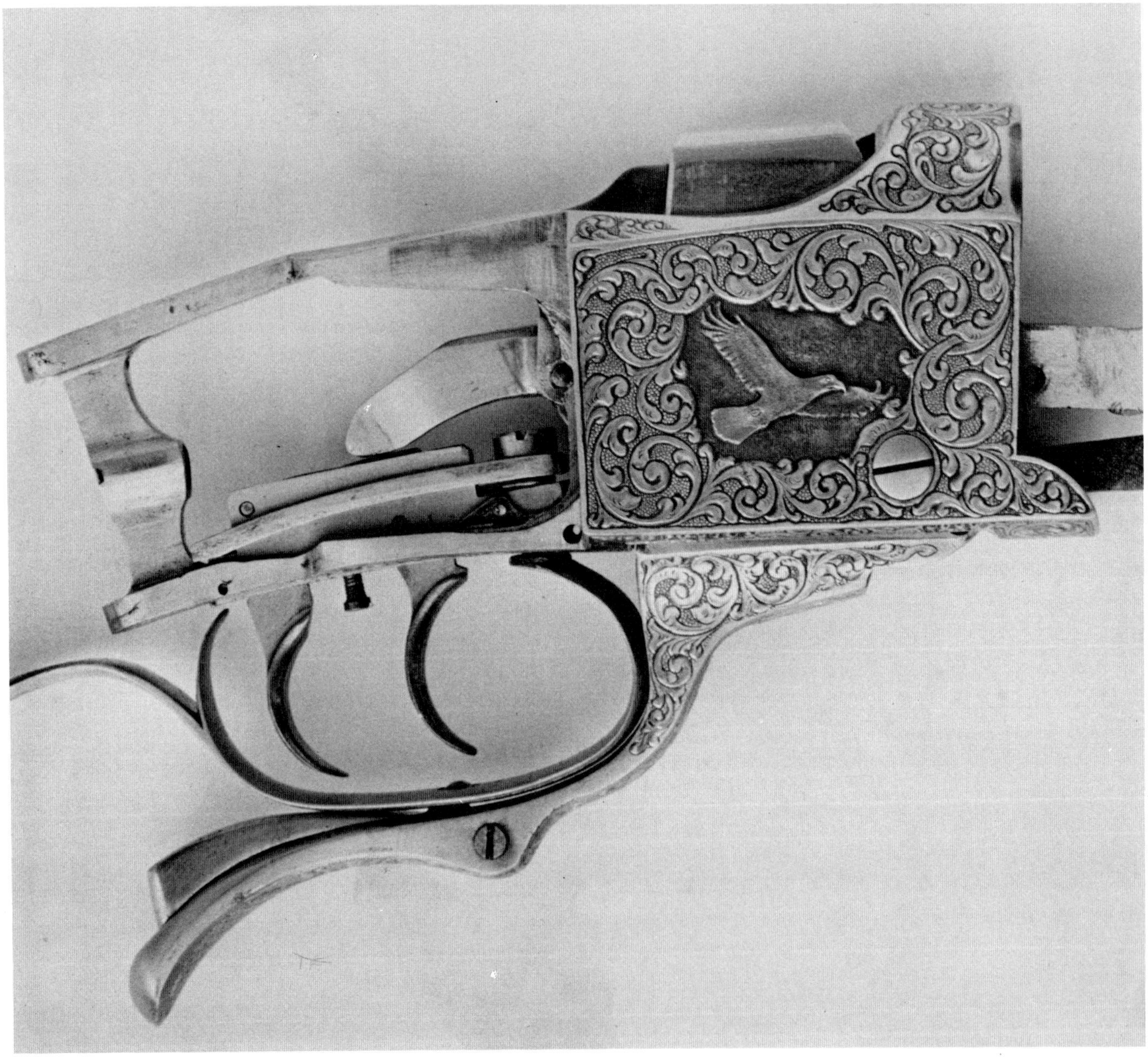

**Mr. Scott has been working for several months on this Ruger No. 1 in his spare time. The decoration includes "Nimschke" style scroll-work, with beaded background and an eagle in flight sculptured in the steel. Mr. Scott has also added double set triggers and a special safety.** (Photo by Eugene White)

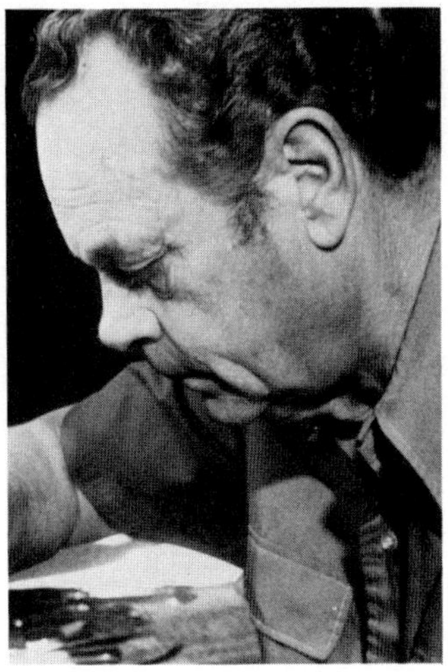

# George Sherwood

*Winchester, Oregon*

A long-time gun enthusiast, in the spring of 1941 George Sherwood went to work for Winchester Repeating Arms Co. where he repaired United States Garand rifles. He enjoyed working for the world-famous arms company and had the opportunity to meet the late George Ulrich, the last in a line of five generations of arms engravers.

Mr. Sherwood learned as much as he could from the engravers at Winchester before he entered the Army in late 1942. During World War II he served as an armament officer in Combat Command B of the 4th Armored Division. His duties took him to France, Belgium, Germany, and Czechoslovakia. When the war ended he was mustered out with the rank of captain.

After the war Mr. Sherwood went to work for the Niedner Rifle Corporation of Dowagiac, Michigan as assistant manager. When the famous custom rifle company closed its doors because of financial difficulties, he went back to school, where he received degrees in industrial relations and law.

In the years that followed, while Mr. Sherwood worked as a corporate executive, he continued engraving as much as possible. Since 1974, when he went into engraving full-time, he has never been happier.

In 1973 he learned about the Gravermeister, which he uses in conjunction with traditional tools. At present he puts in forty to fifty hours per week at his bench.

Mr. Sherwood prefers to work on floor plates and triggerguards, flat-sided repeating shotguns, and knives. In his designs he principally uses several variations of American scrollwork. In addition to private collectors, several knife manufacturers are his clients. They include Western Cutlery, Schrade, and Ka-Bar.

Mr. Sherwood joined the National Rifle Association in 1937 and is a life member. In addition he is a member of the Oregon Arms Collectors Association and the Knife Collectors and Dealers Association, an associate member of the Knifemakers Guild, and a charter member of the Oregon Knife Collectors Club.

Mr. Sherwood is an avid fisherman who travels the waters of the Pacific Northwest in his thirty-seven-foot diesel cabin cruiser. He is also a former ski jumping champion and a Porsche car buff.

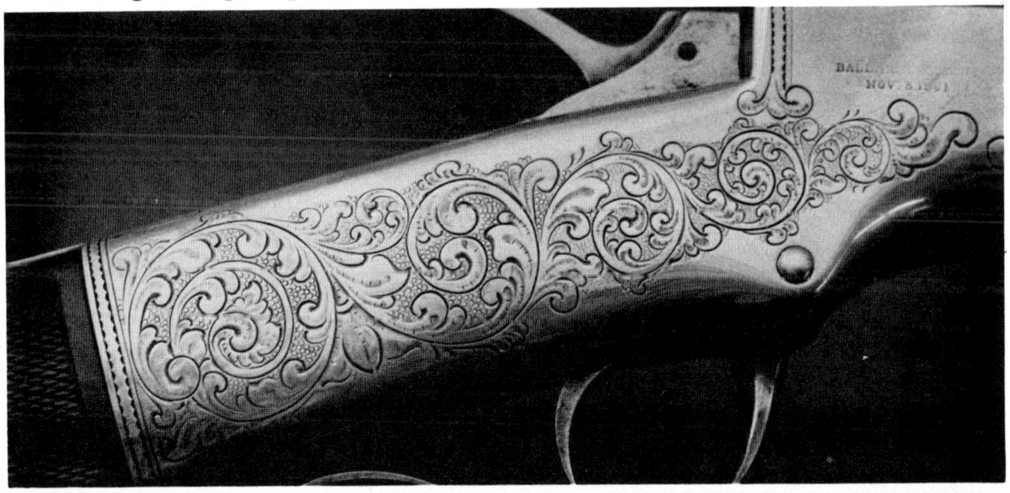

**The receiver of this Ballard rifle has been cut by George Sherwood with a Nimschke style floral scroll and circle-matted background.**

**A versatile engraver, Mr. Sherwood works in many styles. Here he has engraved the blade of a knife handmade by Bill Duff with relief Germanic scroll and a relief steel ram's head.**

Two views of a Savage 99 lever action rifle used by Mr. Sherwood as a display piece. Each game scene is cut in modeled relief with flat relief scrollwork. The receiver has been finished gray.

Bas-relief American style scroll has been used in conjunction with a silver overlay. Mr. Sherwood obtains many of his overlays from Sid Bell.

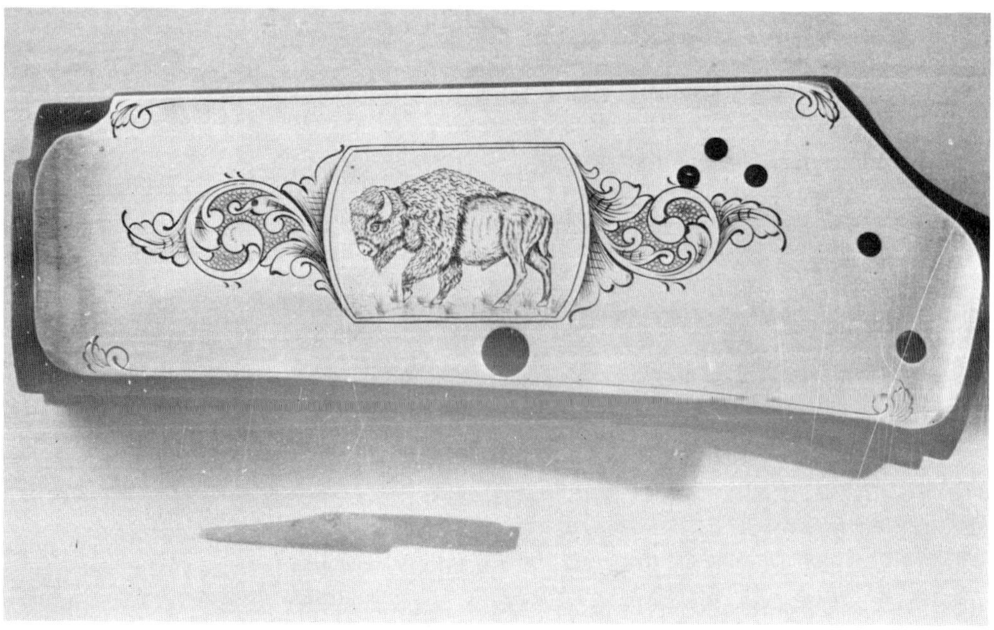

A Marlin receiver is shown here with a line cut bison and American scroll. The object below the receiver is a graver.

Born in Louisville, Kentucky, Ben Shostle is a former tool and die maker who left Chrysler Corporation in 1967 to enter the firearms business. He is the founder and owner of "The Gun Room," a retail gun shop located in Muncie, Indiana. In spite of the demands of his retail business, most of Ben's time is devoted to arms engraving.

As with many American-born engravers Ben is self-taught. He makes use of the Gravermeister, as well as the hammer and chisel method for his work, and estimates he has engraved some 500 guns.

Ben's other interests include deep-sea fishing, hunting, and his work with the Shriners; he has attained the thirty-second degree of Masonry.

In addition to engraving, Ben is also skilled in stock-making, carving, and gunsmithing. He no longer accepts repair or stock work, however, because of the demands on his time for engraving.

# Ben Shostle

*Muncie, Indiana*

**The Ruger Old Army percussion revolver is the object of Ben Shostle's decoration, using American scroll with punched background. Note the cattle skull motif on the frame ahead of the cylinder.** (Courtesy of the Ramrod Gun Shop)

This Colt Trooper MKIII by Ben Shostle carries nearly full coverage scrollwork and a highly polished finish.

Burt Sinclair has been engraving professionally since 1974. A self-taught engraver, he believes that his experiences outside engraving taught the discipline, patience and perseverance needed for this exacting profession.

His background includes a bachelor's degree from Texas Technological University, where he took courses in the philosophy of art; four years of active duty with the United States Air Force; and fifteen years as a trooper for the Texas Department of Public Safety. Mr. Sinclair feels that his work has also been affected by his association with other engravers and his study of the works of Gustav Young.

Two of Mr. Sinclair's favorite guns for engraving are the Colt Single Action Army and the Colt Government model. He likes the Single Action because of its history and its contours. He enjoys working on the .45 automatic because its flat sides afford a good surface for his designs. His fondness for the Colt automatic also derives from his Air Force experience; as a B26 navigator over Korea, he carried the .45 automatic during sixty missions.

# Burt Sinclair

*Shallowater, Texas*

No two guns engraved by Burt Sinclair are identical unless they are a matched set. His work can be identified by the $ on the frame; there is also a hand-lettered certificate for each engraved piece.

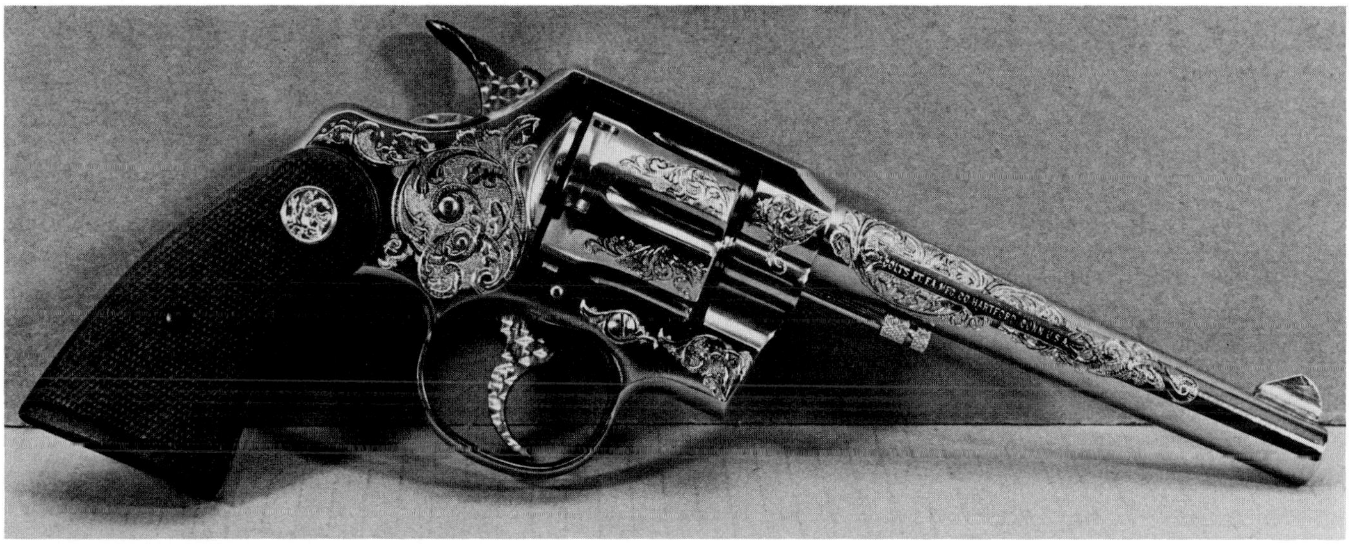

**This Colt exhibits open leaf scrollwork. It is nickel finished.**

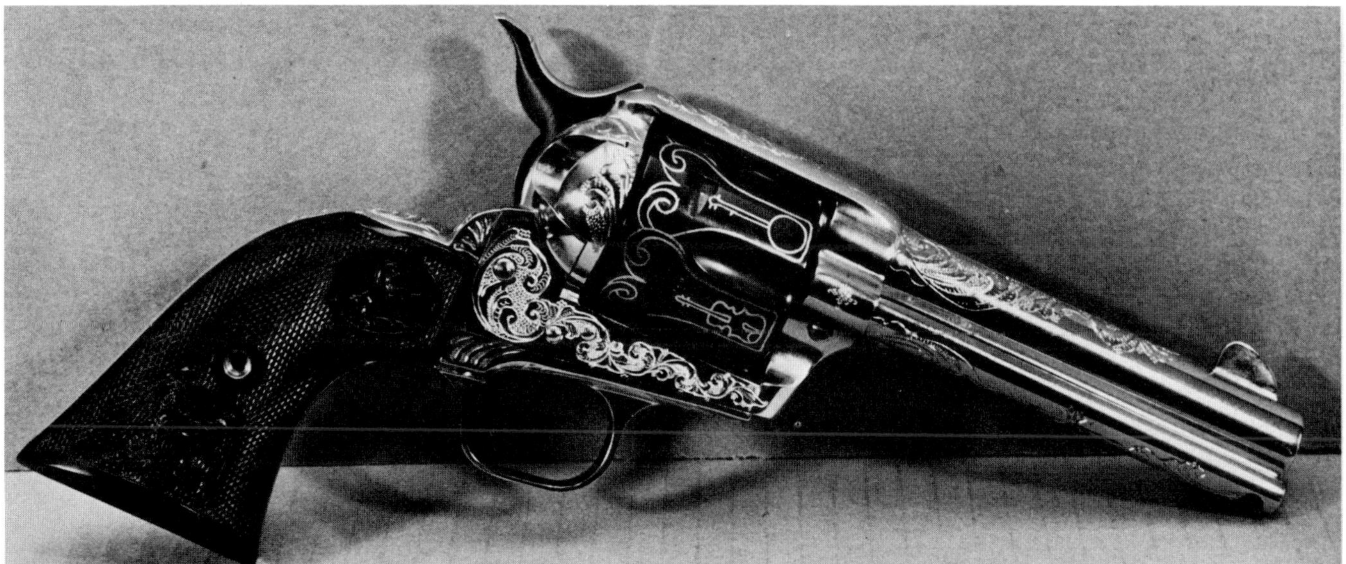

**A gun belonging to the Sinclair family. On the cylinder, musical instruments have been inlaid in gold. The frame is finished in nickel and gold.**

Bold scrollwork is displayed on this Colt Combat Commander.

Three-quarters of this Colt has been engraved. It is nickel finished and has ivory grips.

The engraving on this Colt Combat Commander model exemplifies light bold scrollwork.

# Russ J. Smith

*Westfield, Massachusetts*

As the chief engraver for the Smith & Wesson Firearms Co., Mr. Smith has engraved over 1,600 firearms during his many years at the trade.

Mr. Smith was educated in liberal arts and design at William and Mary College. He also received both layout and design training for the printing business and vocational training in silver and jewelry engraving.

During World War II Mr. Smith served as a staff sergeant and tank commander with the American armored forces. He fought in Africa, Italy, France, and Germany.

As well as an engraver, he is an accomplished scrimshander, preferring to work on whalebone. For engraving he prefers to use the hammer and chisel without the burin.

Mr. Smith's other interests are boating and music.

Sporting a scrimshawed whalebone grip inlay, this Model 52 Smith & Wesson has been engraved by Russ Smith in a style that he calls deep scroll Arabesque. Several American engravers refer to this Germanic style of decoration, which incorporates leaves with scroll spirals, as Arabesque; others call it floral scroll and Austrian engravers call it altdeutscher Laubornament. This confusion over terminology results from this country's lack of a school or guild for engravers.

German oak leaf scroll and game scenes in high relief have been engraved on this partially assembled Smith & Wesson by Russ Smith.

Russ Smith has engraved a wide variety of designs on this plate, which is used as a display piece by the engraver. Note the eagle is inlaid in copper.

This Smith & Wesson sideplate displays the great seal of the United States inlaid in 18 karat gold and silver by Russ Smith.

# George B. Spring III

*Salem, Connecticut*

George Spring is a second-generation engraver whose father was a jewelry engraver and clock repairman for many years. Though Mr. Spring is mostly self-taught in firearms engraving, he learned the technique of inlaying from his father.

Four years at a technical school in Middletown, Connecticut prepared Mr. Spring for work in the machine tool industry, which led to his certification as a journeyman tool and die maker. He first worked in his uncle's machine shop, and then for the next four years he was employed in the tool room of the famous Pratt and Whitney Aircraft Corporation.

When Mr. Spring's ability as an engraver improved, he left the machine tool business and went to work as an engraver in the Pederson division of Mossburg Firearms. He worked there from 1973 until Pederson ceased operation in 1975. His skill and talent easily found employment next with Colt Industries in their custom gun shop, where he worked until June 1977.

For the last two years Mr. Spring has conducted his own engraving business in Salem, Connecticut. In addition to firearms and knives, his skills are applied on presentation plaques and ship bells. Recently he's concentrated on making engraved and cased pistol sets, which he will offer for sale at a later date.

Like many Americans of his age, Mr. Spring's activities were interrupted by the war in Viet Nam. He served as an air crewman in the Navy from 1963 till 1967, attaining the rank of third class petty officer and serving two tours of duty in the combat zone.

Counted among his clients are the Sultan of Oman and entertainer Jerry Lewis, who has a golf club engraved by Spring. He has also been called upon to engrave a case lid inscription for President Carter.

Mr. Spring feels, as do many contemporary engravers, that it is a privilege to be one of the few people in America who follow in the footsteps of the great master engravers of times past. Of the nineteenth century artists he considers Gustave Young (Jung) the greatest, a man whose works will be treasured for years to come.

**Two views of a Ljutic single barrel trap gun by George Spring. The scrollwork is in high relief and the figures are in raised gold inlay. Note the flying clay target inlaid below the serial number. When the gun is blued, gold-inlaid birdshot is seen to be chasing the target.** (Photo by G. Allan Brown)

**A partially disassembled Perazzi over/under shotgun engraved in high relief by George Spring. Note his tasteful use of oak-leaf decor with scrollwork, and the unique, slightly geometric designs he has incorporated into the scroll.** (Photo by G. Allan Brown)

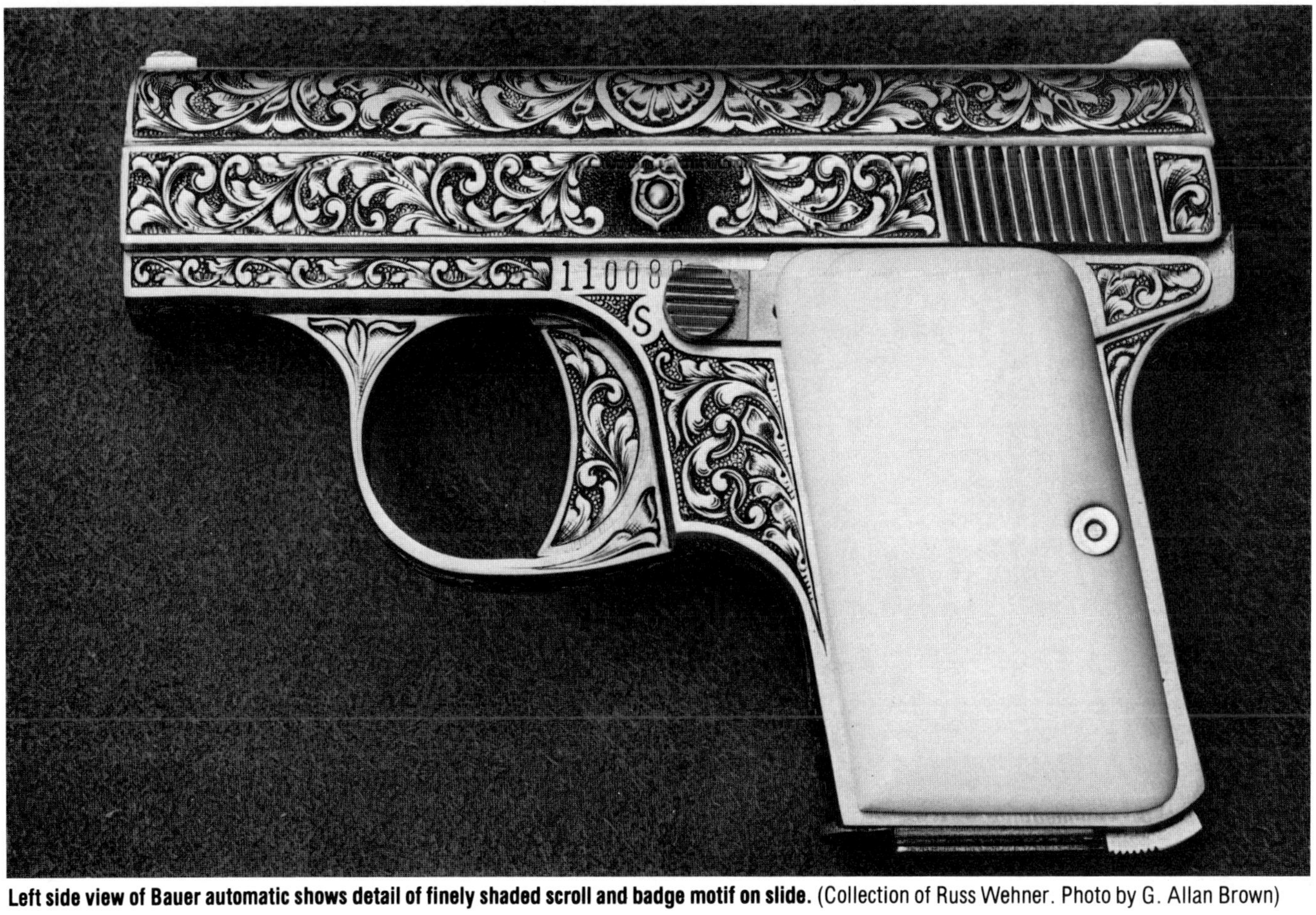

**Left side view of Bauer automatic shows detail of finely shaded scroll and badge motif on slide.** (Collection of Russ Wehner. Photo by G. Allan Brown)

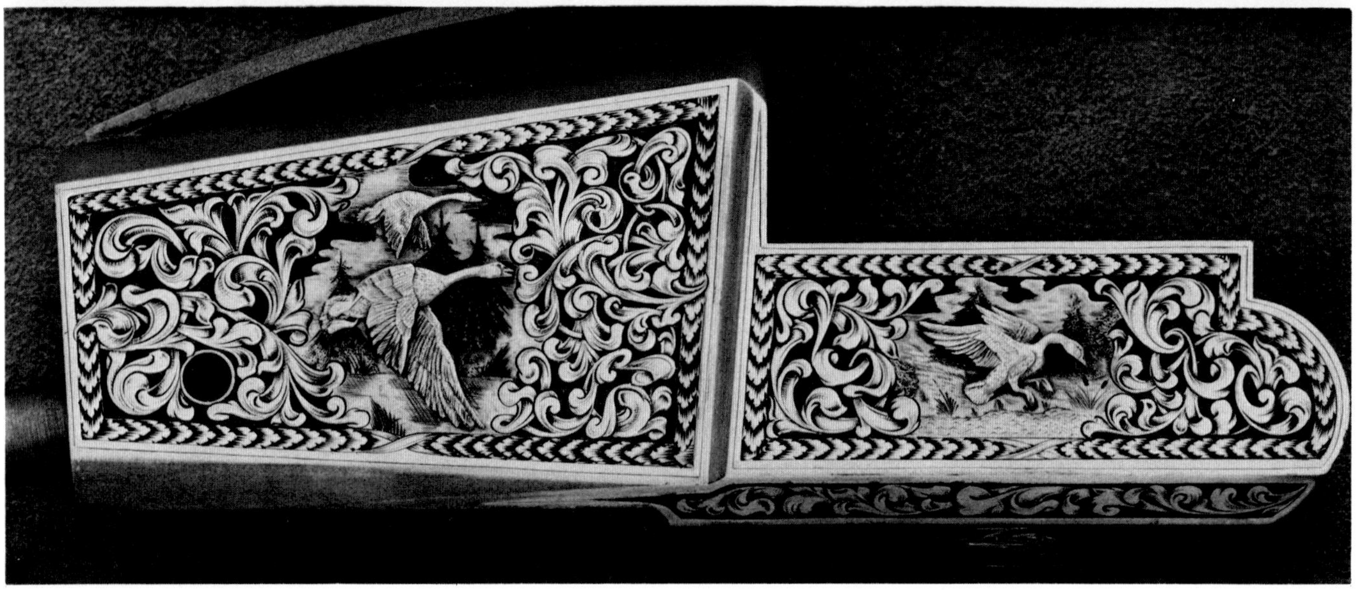

**Shown in the "white" for clarity of detail, these three views of a Ljutic Mono Gun (Trap) illustrate Mr. Spring's unique style of precisely shaded scrollwork. Note how some of the leaves wrap themselves around the wheat chaff borders to increase the three-dimensional effect. All seven figures, including the serial number, are inlaid in raised modeled gold. Note Mr. Spring's signature on the left side.** (Collection of Brent Umberger. Photo by G. Allan Brown)

A graduate of the gunsmithing school at the Colorado School of Trades, Mr. Swartley has been involved with engraving and firearms for over twenty years. Shortly after his graduation from gunsmith school, he was called for two years active duty with the United States Army. He served at Fort Eustis, Virginia, where he was a member of the Army rifle team. Fortunately his engraving skills were not dulled by military service because he kept his tools handy for occasional work.

# Robert D. Swartley

*Napa, California*

After military service Mr. Swartley was employed by the prestigious firm, Griffin and Howe of New York City, which was the gunsmithing division of Abercrombie and Fitch Sporting Goods. While there he had the rare opportunity to work with the finest engraver of his day, Josef Fugger. As a young man, Mr. Fugger (who is now retired) had benefited from working with the late Rudolph Kornbrath (1882-1946), who was the finest engraver of his time.

After two years with Griffin and Howe (1962-1964), Mr. Swartley became a free-lance engraver. He now lives and works in Napa, California, where he engraves firearms and steel dies from which he makes limited edition wildlife prints. He prefers to engrave guns with the burin (hand graver) and the Gravermeister apparatus. At present he prefers to work only for private individuals, who usually accord him more artistic freedom.

**Profuse coverage with Mr. Swartley's unique scroll in high relief and raised-relief animals inlaid in gold puts this Smith & Wesson in the category of an exhibition piece. This piece was displayed on the cover of the April 1978** American Rifleman.

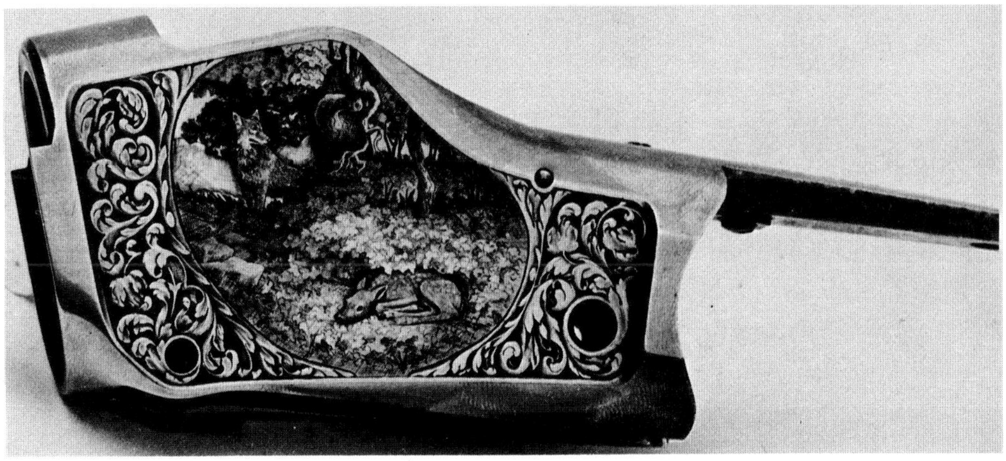

**Mr. Swartley's unique style of leaf decoration is pictured here surrounding a bank note style game scene. This piece was also featured in** The Book of Winchester Engraving **by R. L. Wilson.**

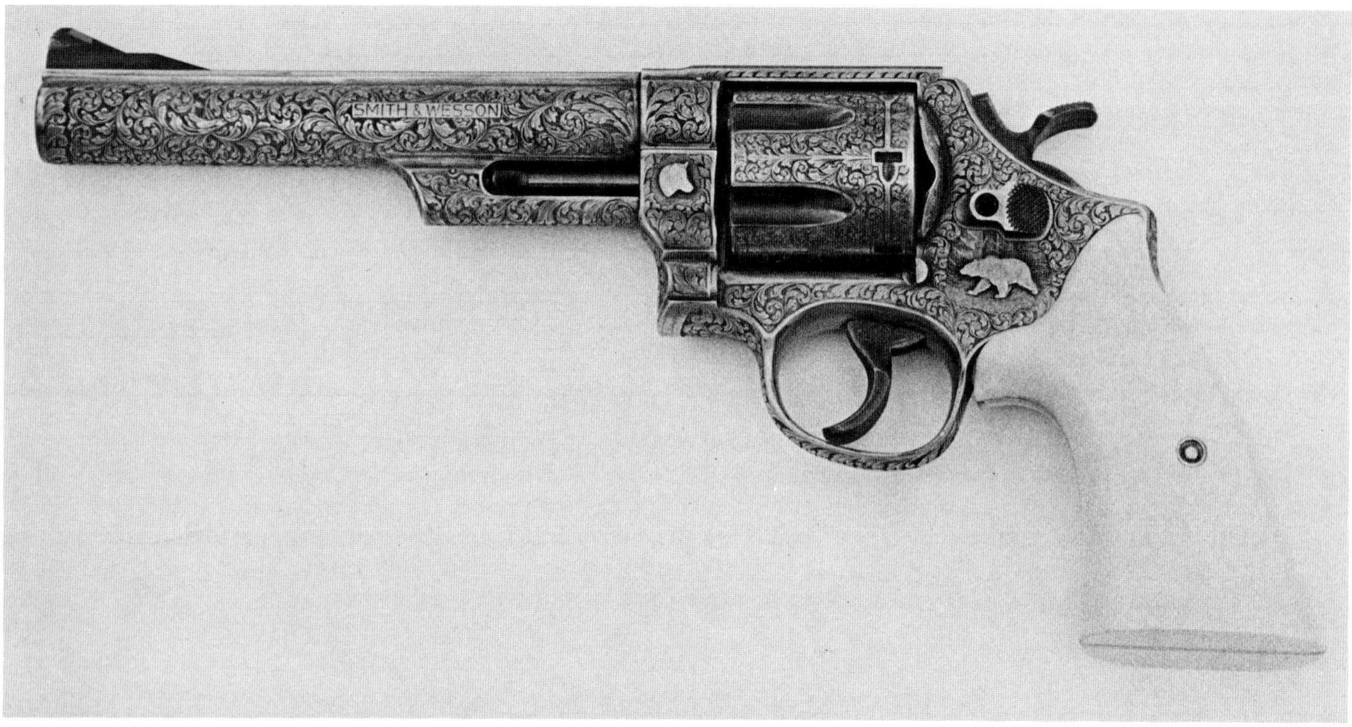


Finished in French gray and highlighted with raised gold inlays, this Smith & Wesson Model 29 features extra-full coverage scrollwork done in Robert Swartley's own distinctive style.

This selection of three floor plate-triggerguard assemblies illustrates the variety of styles Mr. Swartley can execute. Minute detail is characteristic of his work.

Two views showing the detail of Robert Swartley's scroll design, as well as his fine line scenes. Note that the central figures of each scene are in flat relief with bank note style detail.

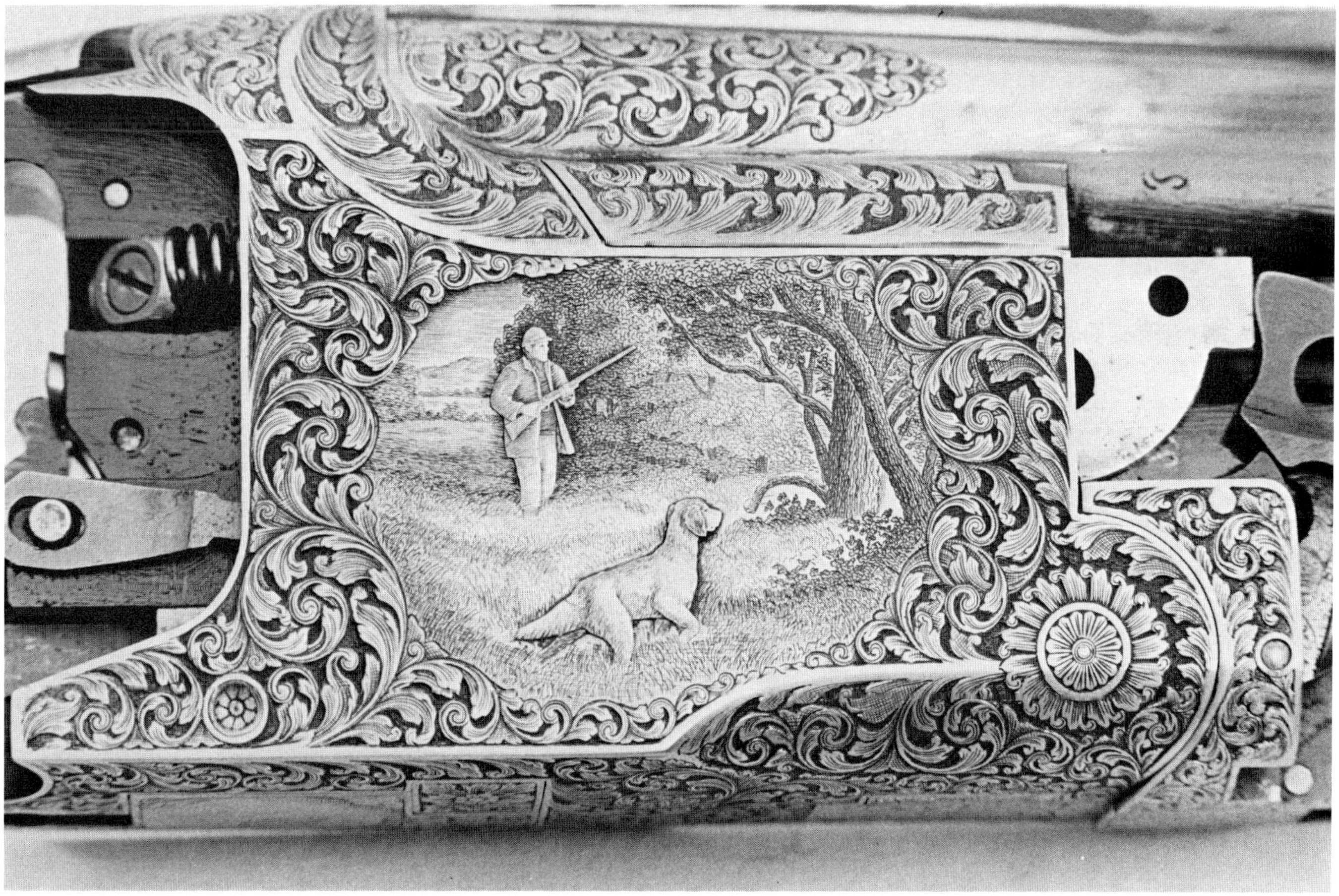

Another example of Mr. Swartley's fine detail work. Once again the central figure is in flat relief with bank note style detail.

Mr. Swartzfeger is one of several Americans who use their engraving skill as a sideline employment and a creative outlet. He is entirely self-taught and is quick to credit *The Art of Engraving* by James B. Meek as his major source of arms engraving information.

A native of Washington, Pennsylvania, Mr. Swartzfeger has been employed in the local steel industry for several years. His background includes five years in the United States Navy as a medical corpsman and work as a lab technician.

Aside from engraving he enjoys metal, wood, and stone sculpture. He does not limit his engraving to firearms, but also works on such objects as knives, belt buckles, tomahawks, and jewelry. He is one of several engravers who uses the Gravermeister in conjunction with hand tools.

# Monk Swartzfeger

*Washington, Pennsylvania*

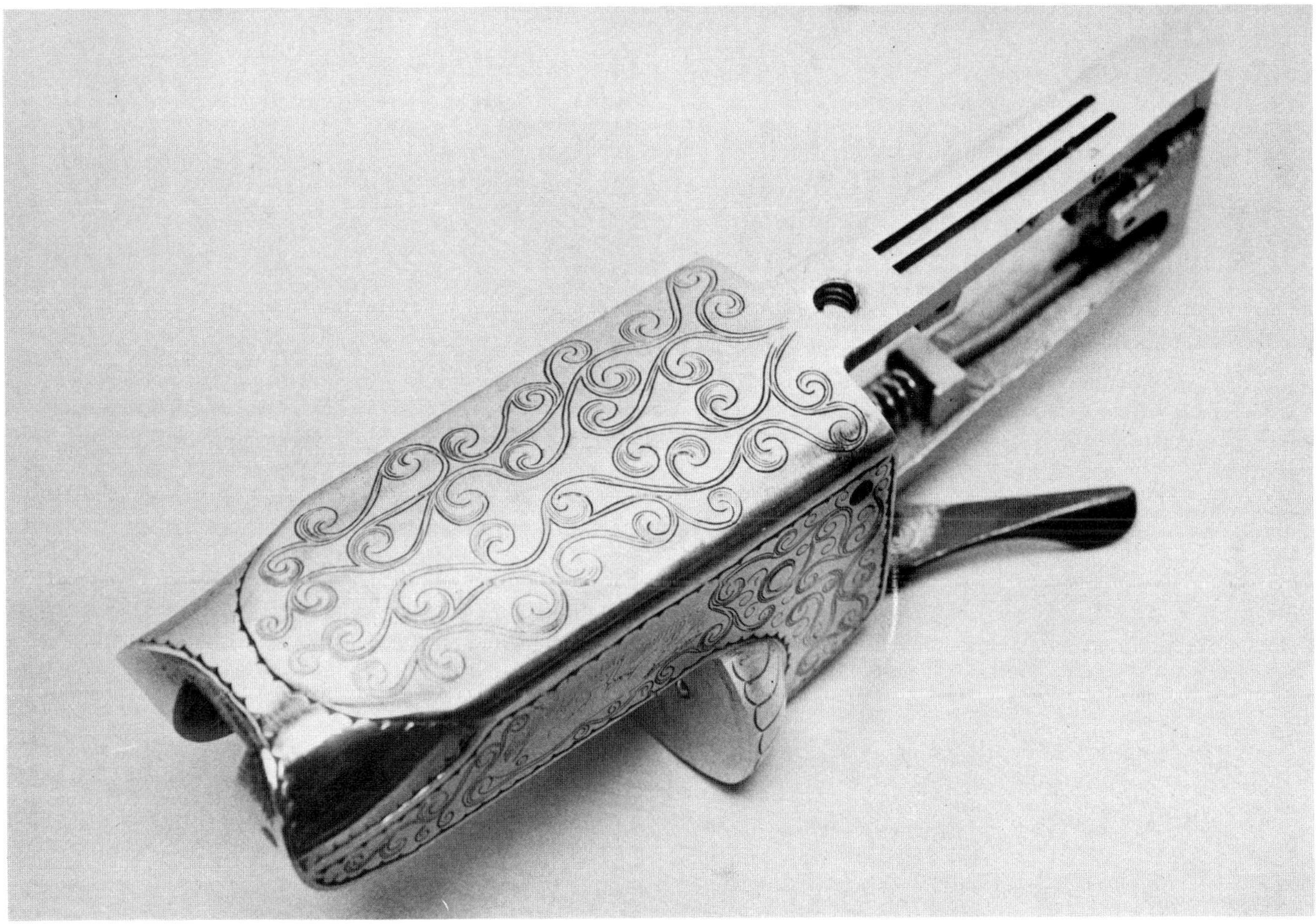

**The action of this Western Field side-by-side shotgun has been engraved in what Mr. Swartzfeger calls "elongated scroll."** (Collection of Buck Shumaker. Photo by J. R. Crouch)

**This floor plate illustrates Monk Swartzfeger's version of large scrollwork.** (Courtesy of Tom's Gun Shop. Photo by J. R. Crouch)

**Beveled line engraving decorates this patch box cover from the stock of a muzzle-loading rifle engraved by Mr. Swartzfeger.** (Collection of Don Sampson. Photo by J. R. Crouch)

**"Stylized scroll" is how Mr. Swartzfeger describes his design on this floor plate; it is almost modernistic in concept.** (Collection of Tom Moore. Photo by J. R. Crouch)

Born in 1902, Roy Vail is the senior engraver featured in this book. Unlike the other engravers, however, he does not accept commissions for engraving work. Instead Mr. Vail is a builder of custom-made rifles and shotguns, and engraves only those arms that he has built or extensively customized.

During his tenure as a gunmaker/engraver he has studied scrollwork with H. H. Hilton and gold inlays with Kornbrath and Griebel. He is a graduate of the Morrisville Technical and Scientific College and holds a Master of Arts degree.

Among his clients have been such notables as President Eisenhower, Admiral Nimitz, and Clark Gable. Mr. Vail is an honorary member of five historical societies as well as the curator of the Warwick and Orange County historical societies.

# Roy Vail

*Warwick, New York*

---

*Author's Note:* During the production of this book we learned that America's oldest active engraver, Roy Vail, died at his home on December 3, 1979. Mr. Vail is survived by his two sons, Allen and Donald, his daughter Norma, and five grandchildren.

Made in .410 gauge, this over/under shotgun by Roy Vail has been engraved with English scroll.

Roy Vail made this 28 gauge double-barreled shotgun using a box lock action that he embellished with fine English scrollwork. The figures are of raised gold inlay. Note how the flush gold barrel band follows the contour of the side clip.

This detail enlargement of a sidelock engraved by Mr. Vail exemplifies good English scroll in light coverage. The game scene is line cut with slight relief.

# Robert B. Valade

*Cove, Oregon*

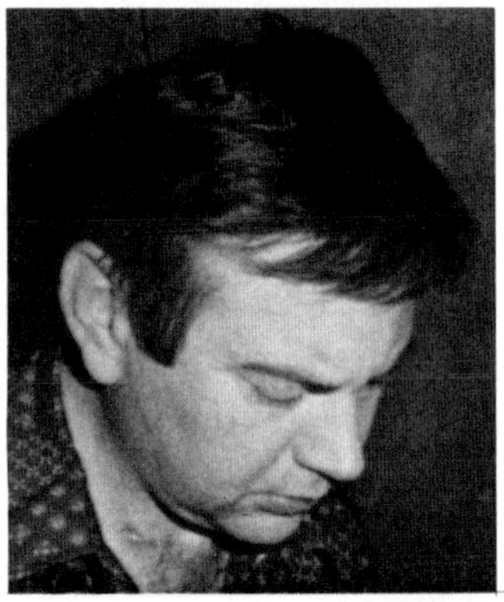

An arms engraver since 1968, Mr. Valade is well-known among knife collectors as a result of his work on Gerber Legendary Blades' special series of presentation knives. He has also plied his craft for other blade makers, including Sid Birt Custom Knives and the Al Mar Knife Company. In all, he has engraved over 1,500 knives of all makes to this date.

In spite of his heavy work load of knife engraving, Mr. Valade has also found time to engrave over 250 firearms during his career. In fact firearms were his main subject until edged weapons began to dominate his time.

In the past Mr. Valade has worked as a hunting and fishing guide in Oregon and still enjoys outdoor activities when time permits.

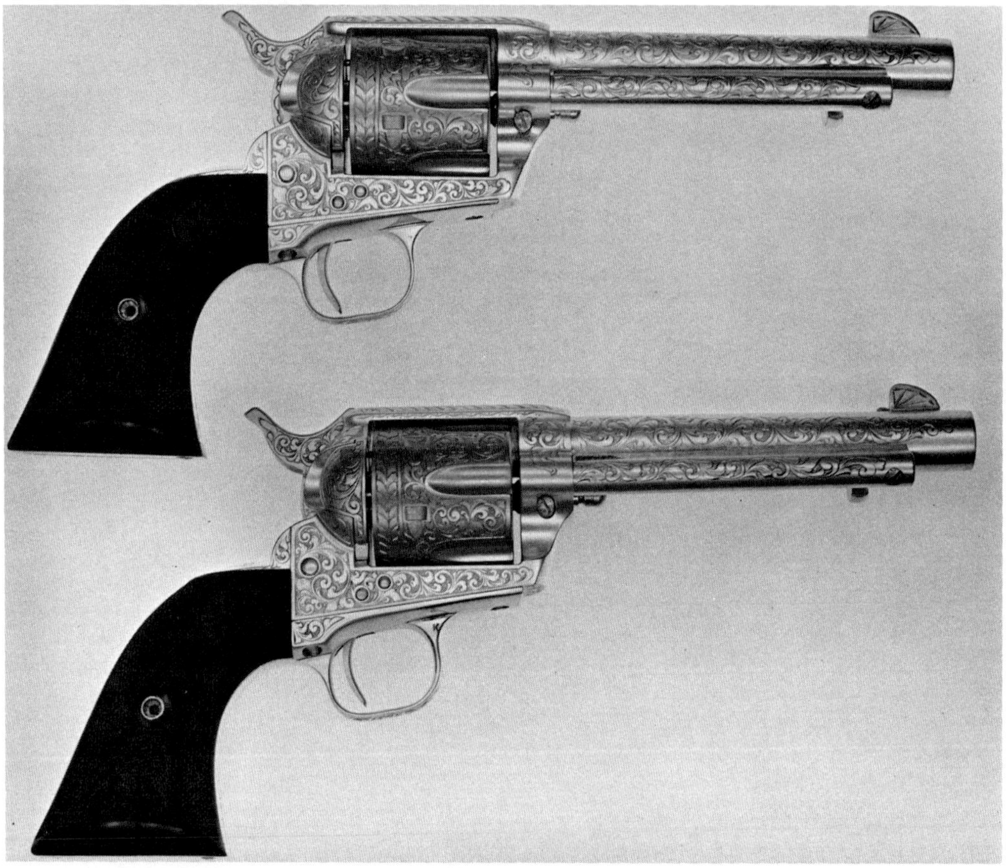

**Several years ago this brace of Colt single action .45s were presented by Colt to famous holster maker Arvo Ojala who brought them to Robert Valade for engraving. Mr. Valade decorated the two pieces with matching bas-relief scroll and a beaded background. The guns have been silver-plated, with gold-plated cylinders.** (Collection of Arvo Ojala. Photo by Ed Prentiss)

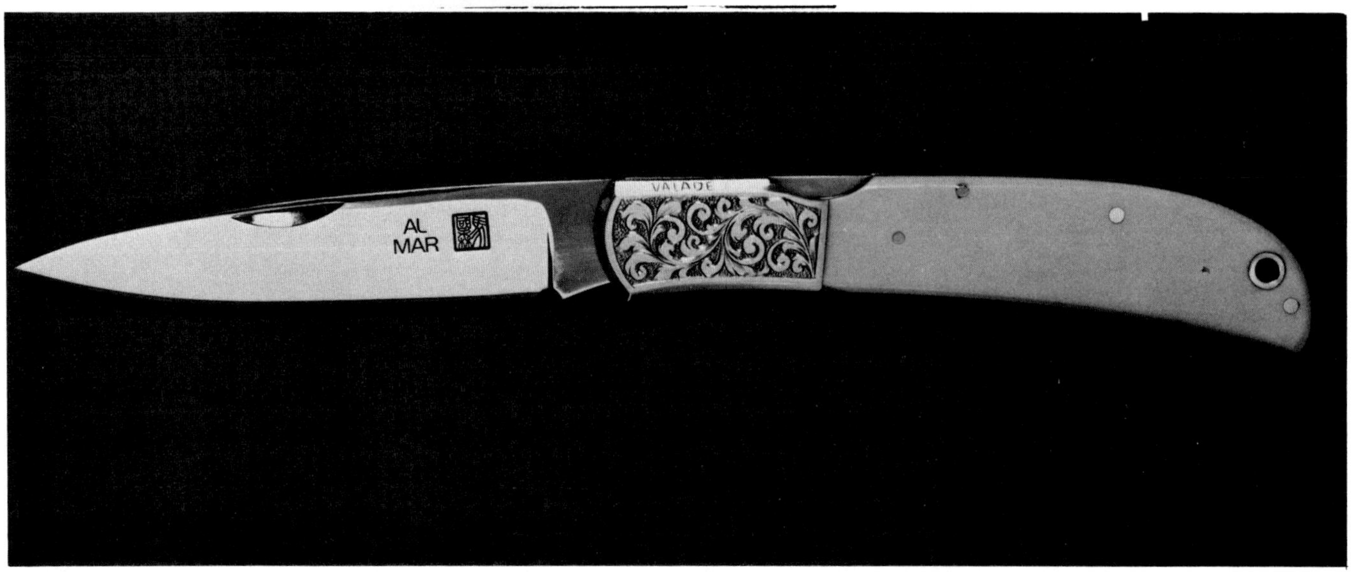


**An Al Mar folder adorned with Mr. Valade's scrollwork. Note the engraver's signature, VALADE, on the bolster.**

# John Vest

*Klamath Falls, Oregon*

John Vest has been engaged in the art of engraving since 1970 when he cut his first practice design. But his interest in firearms goes back much further, for he has done stock-making and carving since his teens. He is, like most Americans, self-taught and credits the *Gun Engraving Review* by Jack Prudhomme and *Art of the Engraver* by J. B. Meek as his main sources of information and inspiration.

Mr. Vest's background includes service with the Oregon Army National Guard, where he attained the rank of sergeant in the infantry and later in the antiaircraft. During his tour he shot with the 6th Army team in the big and small bore events in 1969.

From 1963 to 1977 Mr. Vest worked for the United States Forest Service as a fire management officer in the Winema National Forest. In 1977 he left the Forest Service and began working as a full-time engraver. He works exclusively with hammer and chisel and with the ball type vise. Besides firearms, he also engraves knives, belt buckles, and similar objects.

From 1960 until 1963, Mr. Vest worked as a professional golfer and still enjoys the sport along with hunting, fishing, photography, and western art.

**A Model 40 Smith & Wesson "Centennial" engraved in 1978 by John Vest in borderless relief scroll.**

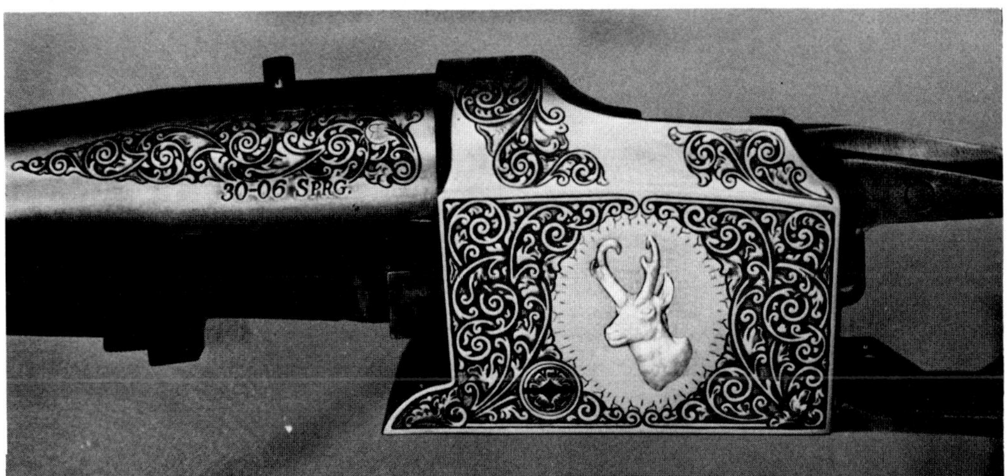


**The Ruger by Mr. Vest is shown prior to bluing in order to illustrate the detail of the scrollwork.** (Collection of Lloyd Jensen. Photo by John Vest)

**This leaping whitetail buck was inlaid on a Sako-Browning floor plate by Mr. Vest. The inlay is 14 karat gold.** (Collection of Mario Zapoznicoff. Photo by John Vest)

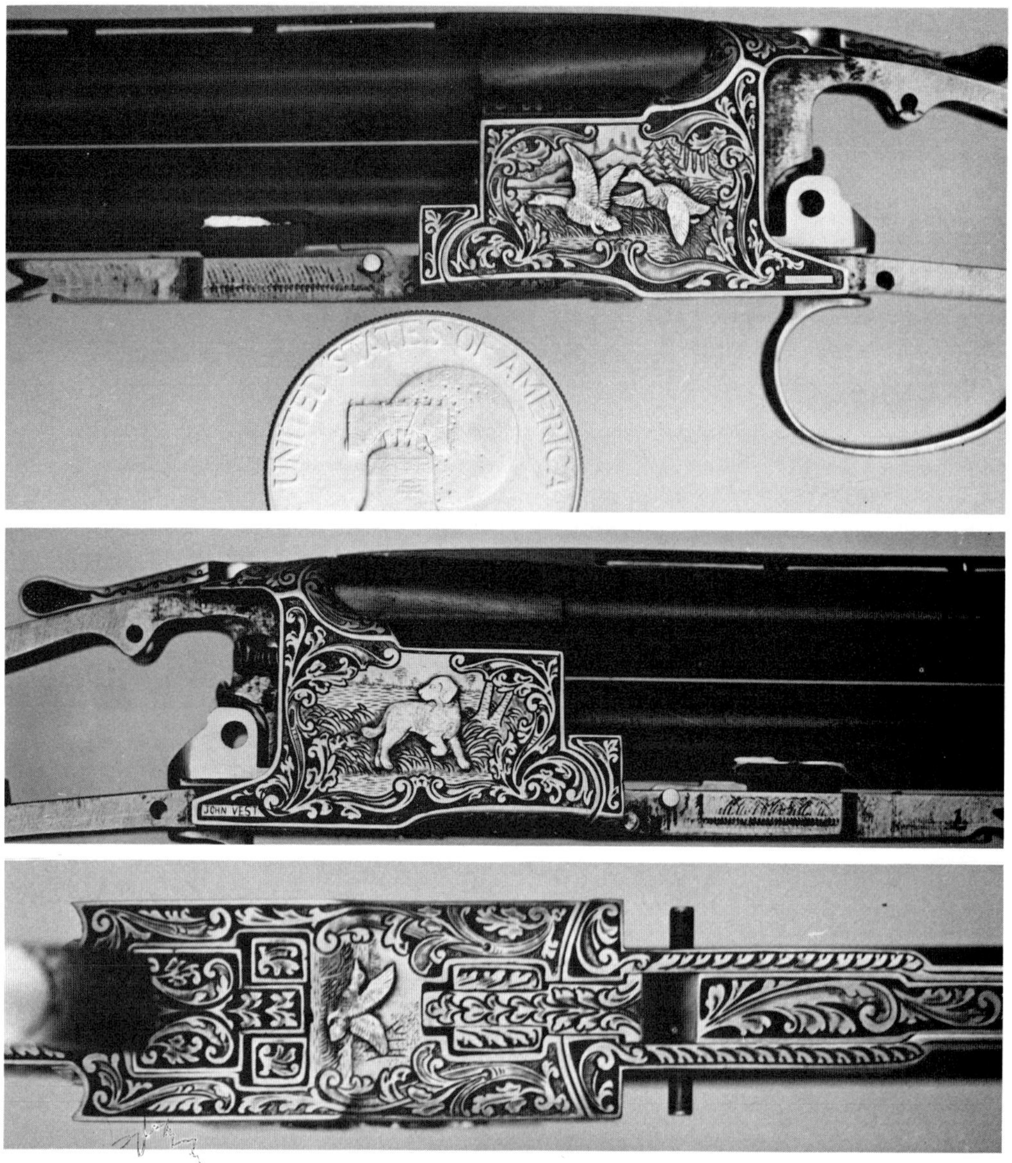


**Three views of a functional** minia-ture **Browning over/under shotgun shown with a silver dollar for size comparison. The entire piece is barely 21 inches long. Handmade by Mike Barrett, it is engraved by John Vest in Browning style P2Q with four 22 karat gold inlays. This piece is in the permanent collection of the Favell Museum of Western Arts and Artifacts, Klamath Falls, Oregon. The Favell collection of miniature guns is acclaimed to be the largest on permanent public display in the world.** (Photos by John Vest)

**Borderless relief scroll accents a lynx head engraved by John Vest into this Smith & Wesson Centennial revolver.** (Photo by John Vest)

Ray Viramontez has served with the United States Air Force for twenty-seven years and presently holds the highest enlisted rank attainable, chief master sergeant. When he retires, Sergeant Viramontez plans to open a full service engraving shop, offering machine and hand engraving for a variety of items.

Mr. Viramontez began his training in 1960 when he was stationed at Wiesbaden, Germany. There he studied under the venerable master engraver Werner Schuck. When he was transferred from Germany in 1962, he ended his apprenticeship. Since then, he has been engraving as a hobby.

Though the Air Force keeps Sergeant Viramontez very busy, he has still found time to engrave hundreds of firearms and knives. His work is well-known in the United States and has been featured in many publications of the arms trade. In spite of his busy schedule, he has willingly taken time to offer advice to a number of other engravers.

# Raymundo (Ray) Viramontez

*Dayton, Ohio*

Mr. Viramontez is the father of two teenage children and enjoys an occasional game of bowling or tennis. He is a member of the National Rifle Association, the National Knife Collectors and Dealers Association and the Ohio Gun Collectors Association, and is an associate member of the Knifemakers Guild.

**Ray Viramontez has created a unique treatment for this government model Colt. The sides have been engraved with profuse scrollwork, while the top and undersides have been left plain to contrast with the engraved surfaces.** (Photo by Ray Viramontez)

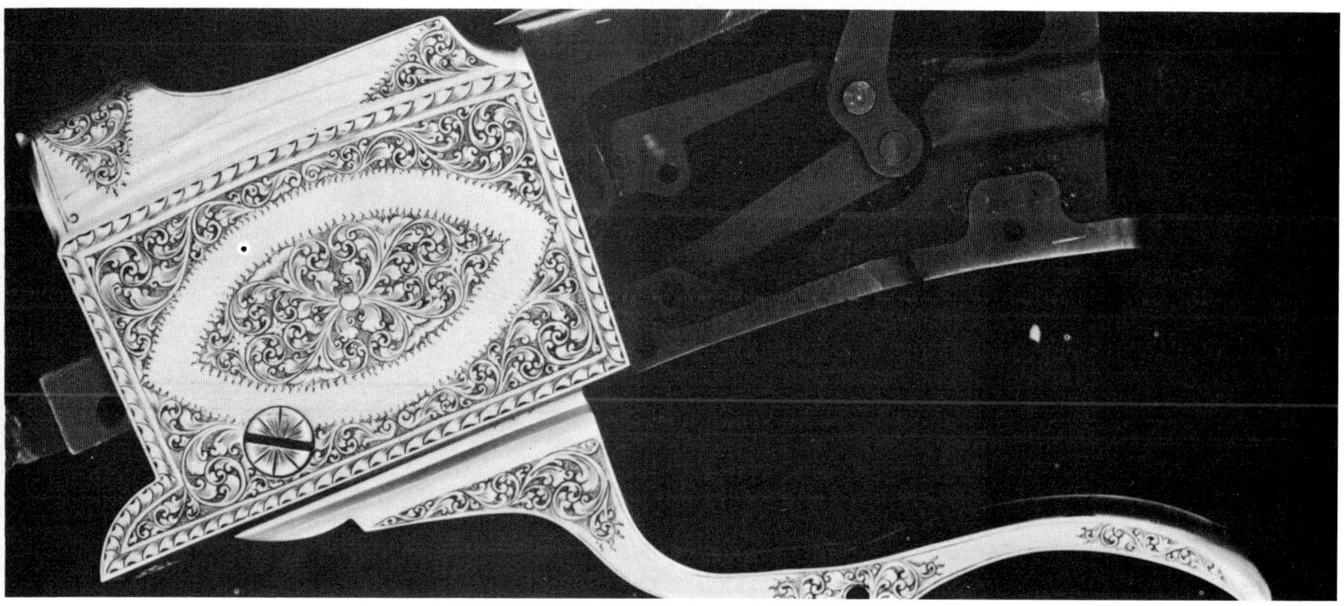

**The receiver of this Ruger No. 1 is part of a rifle presented to** Gun Digest **editor emeritus John T. Amber. The finish is armaloy. Mr. Amber's initials are on the reverse side.** (Photo by Ray Viramontez)

**A pair of custom-made folding knives by Lofgren, embellished by Ray Viramontez.**

**Mr. Viramontez has used German scrollwork to form a cartouche for a relief-engraved ram on this Ruger No. 1 rifle.** (Collection of Ron English. Photo by Ray Viramontez)

**A view of a Winchester Model 12, decorated by Mr. Viramontez with scroll bordering and a cartouche of oak leaves surrounding 24 karat gold-inlaid game scenes.** (Photo by Ray Viramontez)

Vern Wagoner first worked on a firearm in 1955 for an uncle who was willing to allow the watchmaker/jewelry engraver to try his hand at engraving a gun. It was eleven years and many free jobs later before Mr. Wagoner felt justified in calling himself a professional and charging for his work.

After World War II, Mr. Wagoner was mustered out of the Air Corps and entered the Chicago School of Watchmaking. For three years he learned the skills of watchmaking, jewelry repair, and engraving. At the same time he studied at the Art Institute of Chicago, attending evening classes.

In 1963, the Wagoners moved to California for a year, and Vernon furthered his art studies at the Grant Beach School of Arts in Los Angeles. Returning to his home in Waterville, Ohio in 1965, he continued to practice the three crafts of watch repair, jewelry, and gun engraving until 1974.

The Wagoners were interested in relocating in a warmer climate, so in 1974 they moved to Arizona where Mr. Wagoner has practiced arms engraving as a full-time

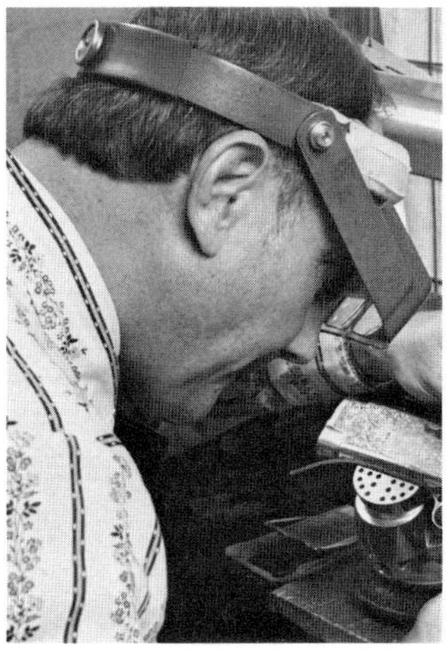

# Vernon G. Wagoner

*Mesa, Arizona*

occupation ever since. He has worked for clients located throughout the United States as well as in France and Guam.

Among his interests are nature study, show pigeons, and fishing. He is also a member of the Masonic Order.

Enlarged detail shows Mr. Wagoner's scrollwork, as well as a covered wagon scene on the cylinder of this Ruger "Old Army" percussion revolver. Note the signature below the serial number; it reads: ENG. BY VERN WAGONER.

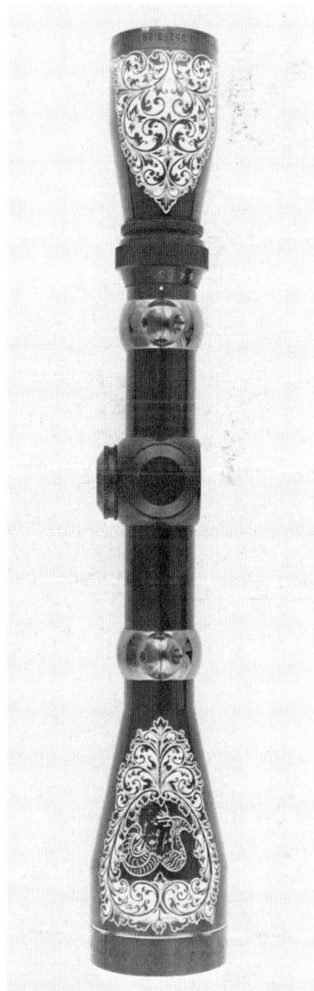

**Something rarely seen in arms decoration is an engraved rifle scope. Mr. Wagoner has engraved through the black anodizing of this piece to form German scrollwork and a line engraved figure.** (Collection of Hal Jankofsky. Photo by Bill McLemore)

**The scrollwork on this Smith & Wesson "Highway Patrolman" has been engraved through the original blue finish. Silver wire in the form of spirals has been inlaid on the barrel, cylinder, and frame. The figure of a bear is also inlaid in silver below the Smith & Wesson logo.** (Collection of Marty Mandall. Photo by Bill McLemore)

**Mr. Wagoner combined his skill with that of scrimshaw artist Bill Shotts of Santa Monica, California on this Smith & Wesson "Chief's Special." Mr. Wagoner cut through the bluing to form American scroll and inlaid a figure in the center of the scroll pattern.** (Collection of Jack Mansfield. Photo by Bill McLemore)

# Terrill Lee (Terry) Wallace

*Vallejo, California*

Terry Wallace began studying engraving in about 1964. His main source of instruction was the *Gun Engraving Review* by Jack Prudhomme. Only after designing and executing many practice plates did he try engraving a firearm. Nevertheless, when he compared his work to gun masterpieces, he was dissatisfied.

Fortunately Mr. Wallace met Robert Swartley, who is featured in the Prudhomme book, as well as this one. Mr. Swartley pointed out some of Mr. Wallace's problem areas and gave valuable advice for improvements.

Mr. Wallace works with all traditional styles of arms decoration and uses traditional tools, as well as the Gravermeister. His work is held with either the bench or ball vise.

Mr. Wallace has been working thirty to forty hours per week as an engraver. Presently he is converting from part-to full-time work. He is an avid outdoorsman who enjoys hunting and fishing. His memberships include the Moose Lodge, the National Rifle Association and the Hasting's Island Hunting Preserve.

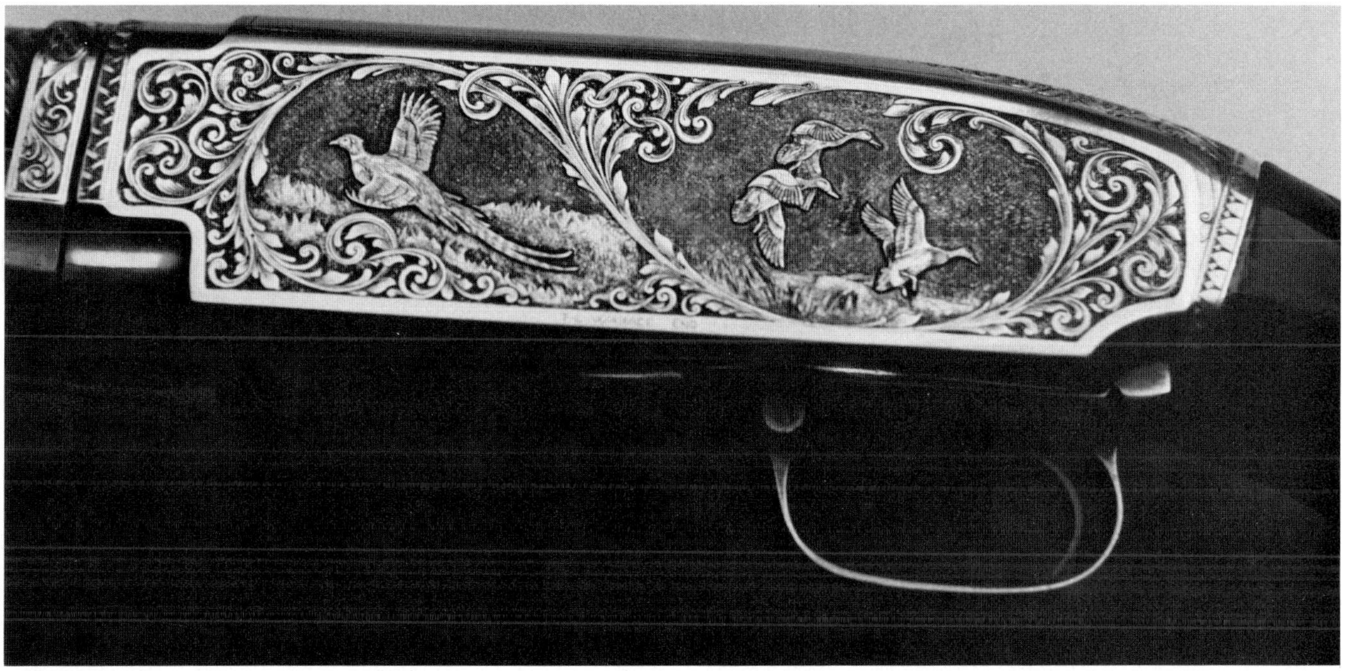

High relief game birds and sweeping scrollwork are used here in these two views of a Model 12 Winchester shotgun. Note the extensive single point shading of the scrollwork's inner leaves.

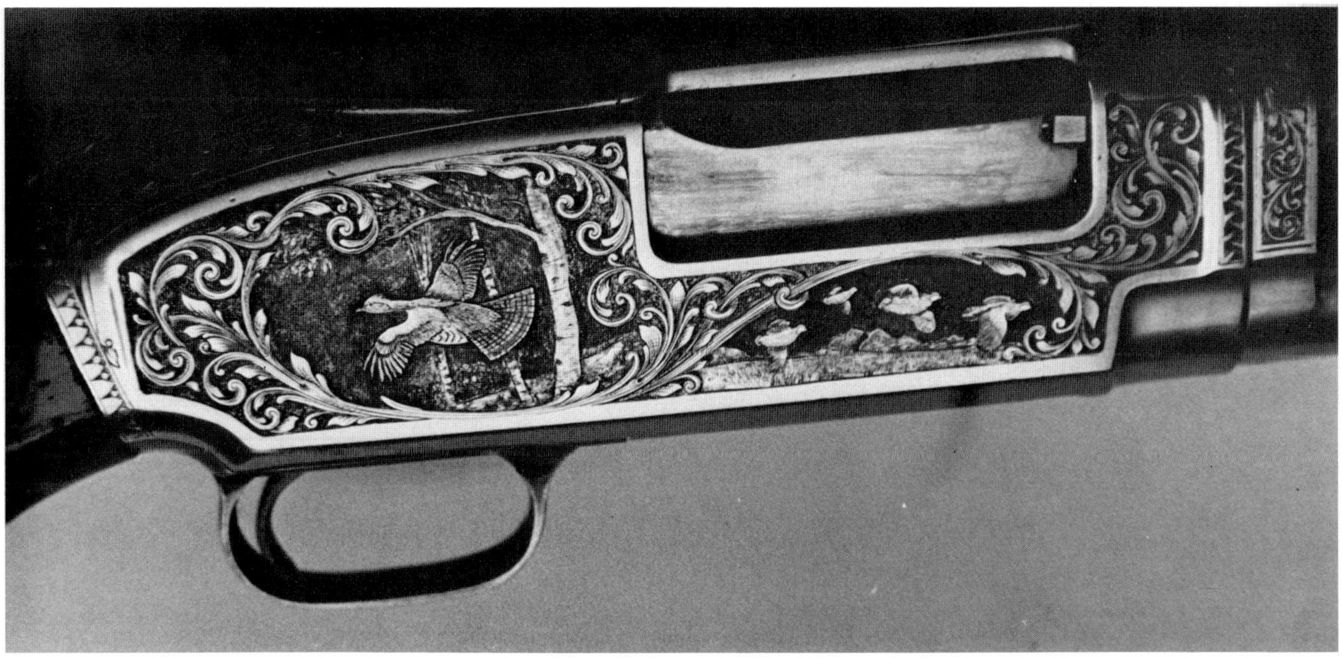

# Floyd E. Warren

*Cortland, Ohio*

Born in 1917 and raised in the small farming village of Kinsman, Ohio, Floyd Warren, like most of America's engravers, expressed himself artistically at an early age. As a child he carved animals from soap bars and made sketches of the farm and wildlife. Many art media were tried by Mr. Warren as he grew up, including wood carving, stone sculpture, cement castings into sand, pastels, oils, pen and ink sketches, and pencil drawings.

Music quickly became his second love and before he was ten years old he was accomplished at the piano, organ, violin, and guitar. Later in 1935 he married the former Belle Baker, whose family of fifteen brothers and sisters were all active country musicians. The family made many public appearances in those days and eventually had their own radio show.

Later Mr. Warren began work at Republic Steel Corporation's Mahoning Valley plant in Warren, Ohio. His employment was interrupted in 1944 when he entered the army, where he served with the engineers as a demolitions man. Returning home after the war, he attended watchmaking and repair school, where he was introduced to the art of engraving.

Though he continued to work at Republic Steel, Mr. Warren was fascinated by engraving and eventually studied the art under Roy Vail (featured earlier in this text) in 1949. Though separated by many miles, Mr. Warren and Mr. Vail were close friends for thirty years.

As an engraver, Mr. Warren uses only the traditional tools. In addition to firearms, he will engrave anything made of metal.

During the past few years he has endeavored to train two apprentices, W. Bush and Tony Tuscano, who he feels have become good artists and engravers.

Aside from music and engraving, Mr. Warren also expresses himself through wood carving, and portrait and landscape painting. Today, most firearms engraving done by Mr. Warren is contracted through John E. Morris of Milton, West Virginia.

**A Colt Python .357 magnum revolver engraved and inlaid by Floyd Warren.** (Collection of Charles Miller. Photo by Terry Vine)

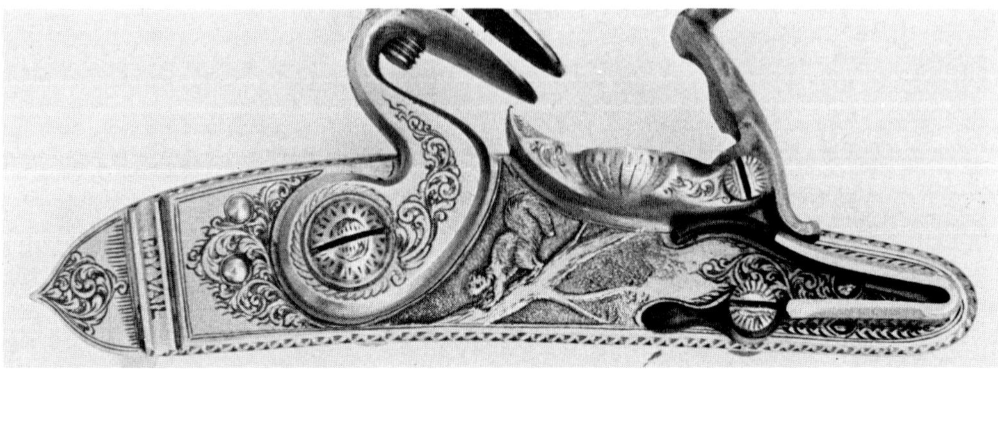

**Engraved for Mr. Warren's long-time friend and mentor Roy Vail, this flintlock features small scroll and a squirrel in relief.** (Photo by Roy Vail)

Featuring nine inlaid gold figures, this Model 42 Winchester sports full coverage German scroll and extensive bordering with inlaid gold wire. Note barely visible signature at the top of the ejection port: ENGRVD BY F.E. WARREN-1975-. (Collection of and photos by Roy Vail)

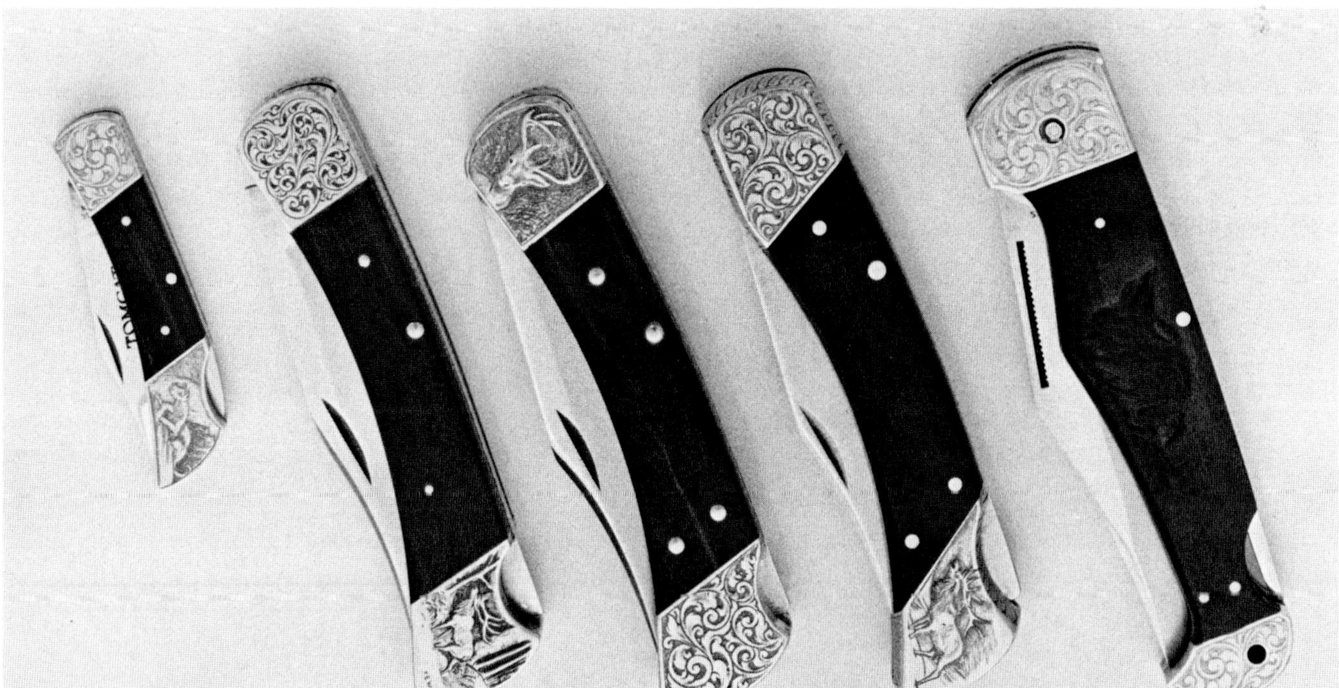

This selection of folding knives is shown courtesy of their owners: Red Adams, William Vassette, Frank Raub, Frank Warren, and David Hackett. (Photo by Terry Vine)

**A Ruger Blackhawk .357 magnum, engraved by Floyd Warren in bas-relief scrollwork. The gun is made of stainless steel, which on this model can be very hard to cut.** (Collection of Dave Hallenback. Photo by Terry Vine)

Born in 1904, Mr. Warren worked for many years in the sign painting trade. He developed skills as a tool and gauge maker during World War II when he worked in an ordnance plant.

Gun work was Mr. Warren's hobby for many years; eventually it became his full-time work. For a number of years he did stockmaking and restoring, as well as engraving, until the demand for his engraving permitted him to drop his other work.

Mr. Warren is self-taught in engraving, which seems to have evolved from his previous experience in art and metal work. He sees engraving as the finishing touch for fine gun work. Using both the hammer and chisel method and the burin for his cutting, he favors the Germanic type of scroll, although he also works in the English and American styles when he feels it appropriate. Gold and silver inlays are usually incorporated into each of his works.

For many years Mr. Warren's work has been pictured in the leading gun publications of this country, including almost every edition of the *Gun Digest* since the early 1960s. All of this has resulted in a great demand for Mr. Warren's talent as an engraver and a tremendous backlog of work. As a result he no longer accepts commissions from new clients and does work only for those long established with him.

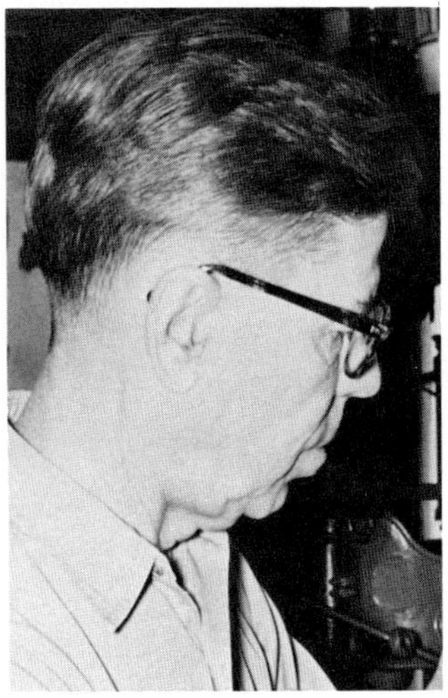

# John E. Warren

*Eastham, Massachusetts*

**Shown here is a Model 1892 Winchester featuring custom stocks and a peep sight. Note the inlay of grape leaves on the side and oak leaves on the bottom. This rifle was featured in the 1974 edition of the** Gun Digest.

**Like this Model 12 Winchester, most of John Warren's better known works feature nearly full coverage German scrollwork, flush gold line inlays, and raised gold figures.**

In 1976 John Warren decorated this .45-60 Model 1876 Winchester with realistic game scenes and scrollwork mixed with an oak-leaf design on the side-plates. Pictures of this piece have been previously printed in the 1977 edition of the *Gun Digest*.

Mr. Warren has depicted a pony express rider, complete with authentic saddle, in raised gold inlay on the frame of this Model 1894 Winchester. This gun was also featured in R. L. Wilson's *Book of Winchester Engraving*.

Sam Welch grew up in Arkansas and attended Ouachita Baptist College where he received a Bachelor of Science in mathematics, with a minor in physics. An additional year of college was acquired in the Air Force meteorology program at the University of Texas, Austin.

While stationed at Turner Air Force Base, Georgia in 1966, he met Ray Viramontez (featured earlier in this book). Having done wood carving, Mr. Welch was very impressed by the work of the engraver. Mr. Viramontez taught him the basics until both men were transferred to separate and distant bases.

Mr. Welch continued studying engraving until 1968, when he returned to civilian life. He was able to pursue his engraving once again in 1974 while he was settled in Alaska. Mr. Welch still lives in Alaska and is employed as the supervisor of the National Weather Service Office at Kodiak.

During 1978 Mr. Welch set up his engraving business and he hopes to be engraving full-time by the end of the year.

# George S. (Sam) Welch

*Kodiak, Alaska*

A member of Elks Lodge 1772, Mr. Welch also enjoys hunting, fishing, boating and shooting, as well as art and photography.

**This enlargement shows the details of Sam Welch's Germanic scroll and oak leaves. Only partially visible is a gold-inlaid grouse on the slide. Note the engraver's signature near the triggerguard.** (Collection of Don Slater. Photo by Sam Welch)

**Mr. Welch has engraved this Model 29 Smith & Wesson with high relief German scrollwork and a raised-relief inlay of platinum, which depicts a rock ptarmigan. The flush gold bordering and barrel inscriptions are not visible when the gun is in the "white."** (Collection of Dave Thacker. Photo by Sam Welch)

**Oak leaves, acorns and a line engraved squirrel are featured on this floor plate by Sam Welch. Wheat chaff borders the bolt holes.** (Photo by Sam Welch)

# Alvin A. White

*Manchester, Connecticut*

An independent craftsman who began his career during the great depression of the late 1920s and early 1930s, Alvin A. White is best known for his elaborately embellished Colt revolvers. But White is also famed for a variety of other work: from scrimshaw to bronze sculpture; from Kentucky rifle making to the design and hand-manufacture of jewelry; from mother-of-pearl inlaying to the restoration of virtually any handmade object of antiquity.

Born in 1915 in Attleboro, Massachusetts, Mr. White is the son of an accomplished die cutter. Since boyhood Mr. White has had an avid interest in guns and shooting, as well as antiques and hand craftsmanship. First trained as a die cutter, during the depression he attended a federally sponsored school for the jewelry trade. While there, he began developing his interest in gun making and engraving. As an air corpsman during World War II, Mr. White was often commissioned by officers to engrave guns or jewelry.

After the war Mr. White returned to Attleboro where he embarked on a free-lance career of engraving, inlaying, and producing just about anything of beauty desired by a client. Except for the period from 1961 to 1975 when he was chief engraver of A. A. White Engravers, Inc., Mr. White has worked independently.

Mr. White sits while engraving and uses a hammer and chisel and a hand graver. He prefers the ball vise and has a small shop with lathe, band saw, and other basic tools. Many of his gravers were handed down from old-timers, including his father.

Mr. White estimates he has done about 2,500 guns in his career, about 25 percent of which are quite deluxe and elaborate. His specialty is relief and flush gold inlays. Owners of his work include several United States presidents and other well-known statesmen, such as King Hussein I, President Anwar Sadat and even Chairman Leonid I. Brehznev.

Mr. White's work has been featured in numerous books and magazines. He has no catalog or price list and is not seeking work.

Four views of the Charter Oak Commemorative matched pair of New Frontier Sheriff's model revolvers, engraved by Mr. White in 1967. The gripstraps are gold-plated; the stocks are of relief carved and gold-inlaid Charter Oak. The gun is profusely scroll engraved with gold inlays; the backstraps have the following inscription: Charter Oak Colts. The case lid is inlaid with a gold plaque inscribed: "Charter Oak Colts"/ MADE EXPRESSLY FOR / DAVID AND JEANNE WOLOCH. The gold-mounted Charter Oak screwdriver, the hinges, case lid plaque, key plate, lock, and silver key were all made by White. The engraver's signature is gold-inlaid on the right side of each frame. This is a most unusual, involved, and distinctive pair of Colts. (Collection of John Snodgrass)

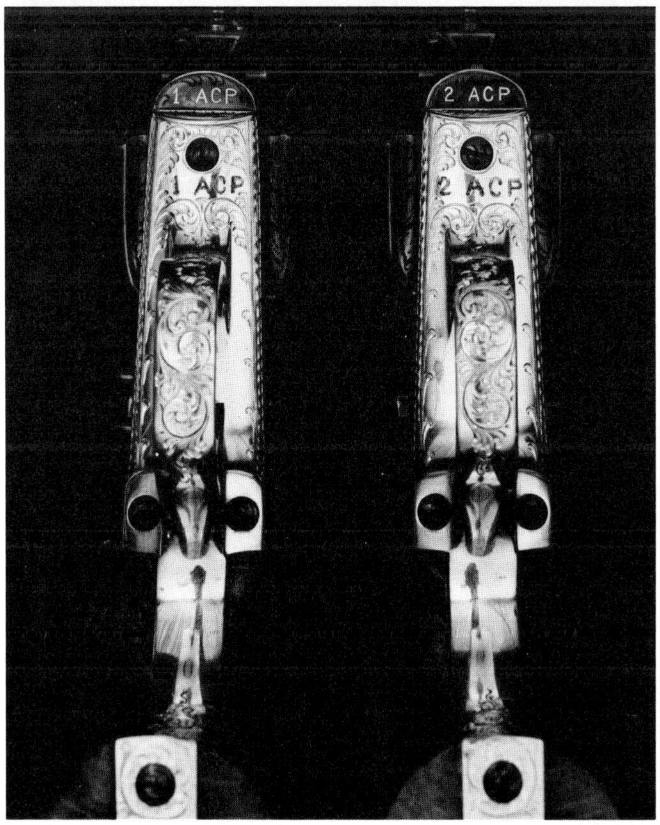


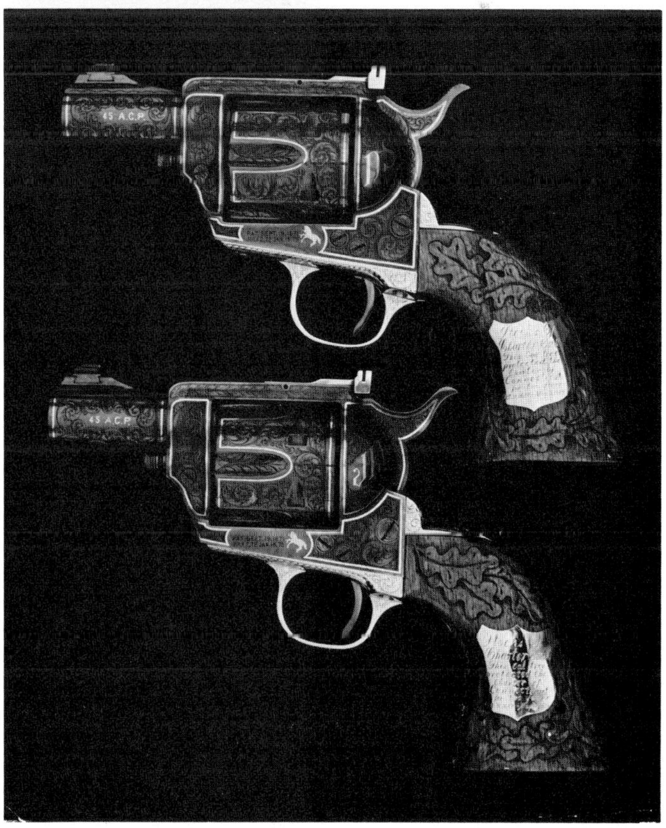

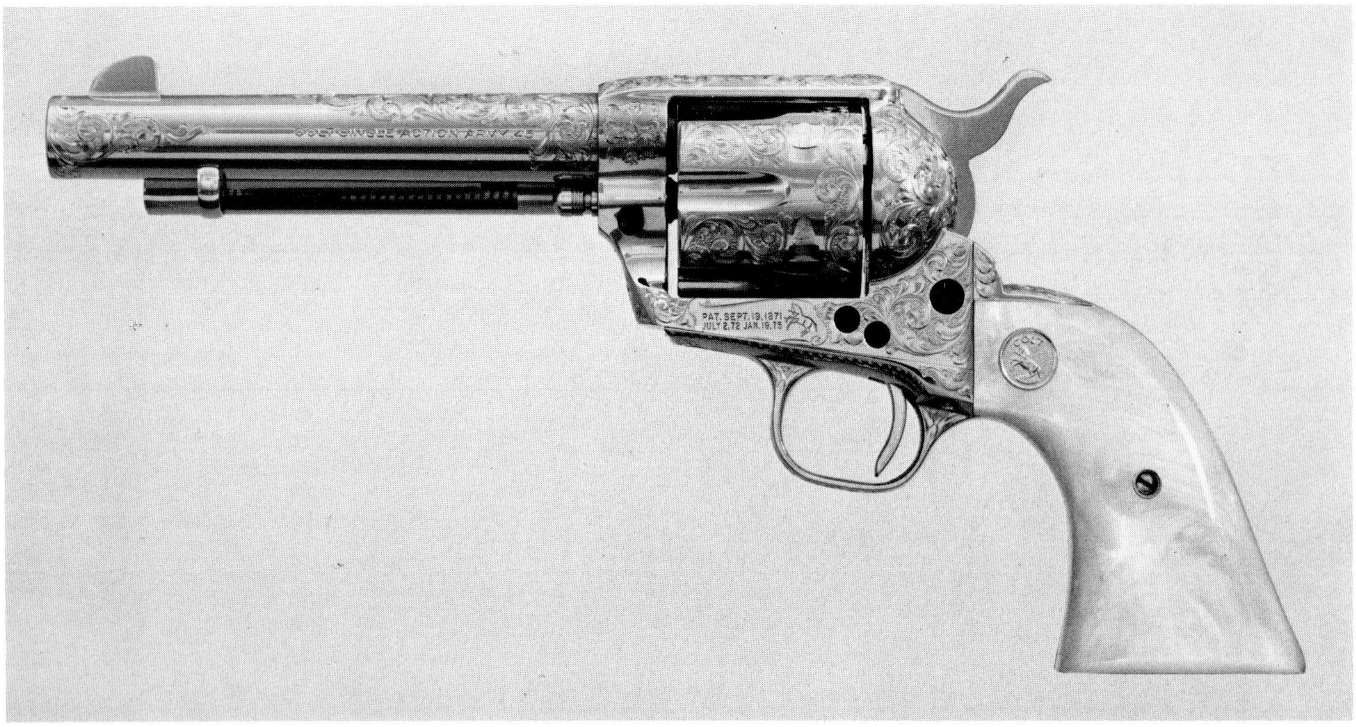

**Example of style C engraving on a Single Action Army made by Mr. White about 1961 as a sample for Colt's factory.** (Courtesy of Colt's Firearms Division)

**Modern Colt Model 1851 Navy engraved by White in an antique style of scroll, with the Colt coat of arms on the left side of the barrel lug, and a sailing ship motif on the right side. The gun has one-piece ivory grips and silver-plated gripstraps.** (Collection of William H. Myers)

**One of the $500 grade World War I commemorative series. The $500 or Custom model pistols, totalling 300 guns, were done by the staff of A. A. White Engravers, Inc. The frames feature gold-inlaid serial numbers, and the frames and slides have a medium amount of scroll coverage; slide scenes and banner inscriptions were hand-cut over prerolled designs. The $1,000 grade pistols (deluxe model) totalled 100 pieces, and Alvin White had an exclusive contract to do the engraving and gold inlay. All bear his gold-inlaid signature cartouche on the receiver beneath the grips. The design features profuse scrollwork, gold-inlaid borders on the slide, gold-inlaid serial numbers, and finely engraved slide scenes and banner inscriptions.** (Collection of W. Gene Moser)

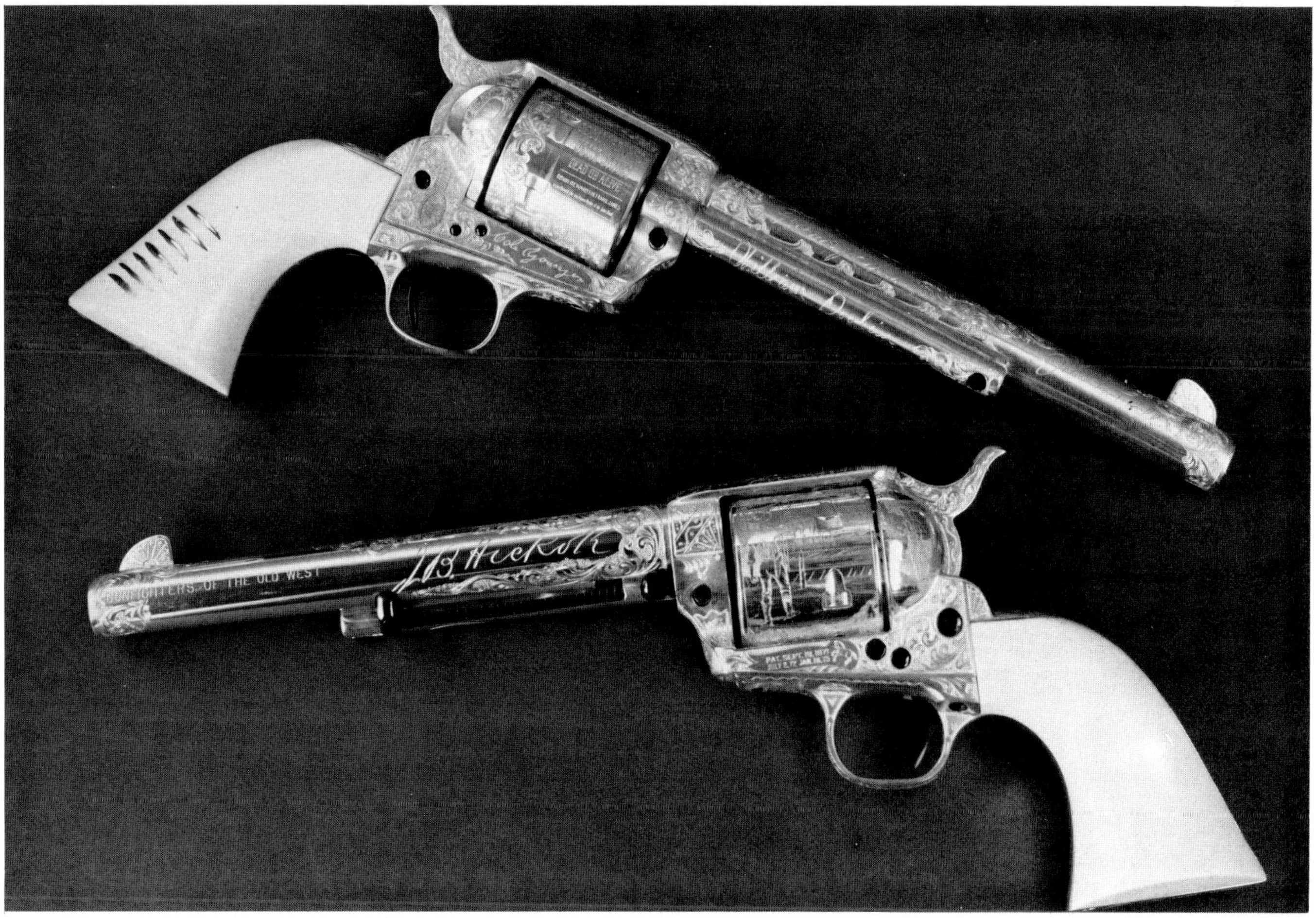


**Two of five deluxe Colt firearms done by White on order of Colt former president William H. Goldbach. The Colt pictured on top is embellished with the theme "Gunfighters of the Old West" and has signatures of Hickok, Garrett, Earp and others engraved in facsimile. It has a Frederic Remington gunfight motif on the cylinder and recoil shield. The matching revolver (below) is inscribed "Outlaws of the Wild West," and is engraved with signatures of Jesse and Frank James, Cole Younger, and other contemporaries. A Frederic Remington print, "Hands Up," is on the cylinder and recoil shield, and a Jesse James wanted poster is on the opposite side of the cylinder. Several pictorial outlaw motifs are cut on various parts. Both revolvers are scroll engraved in Helfricht style and feature ivory grips with notches inlaid in 18 karat gold. The backstraps of both revolvers bear inscriptions of Goldbach's name and title. Both guns are silver-plated.** (Collection of William H. Goldbach)

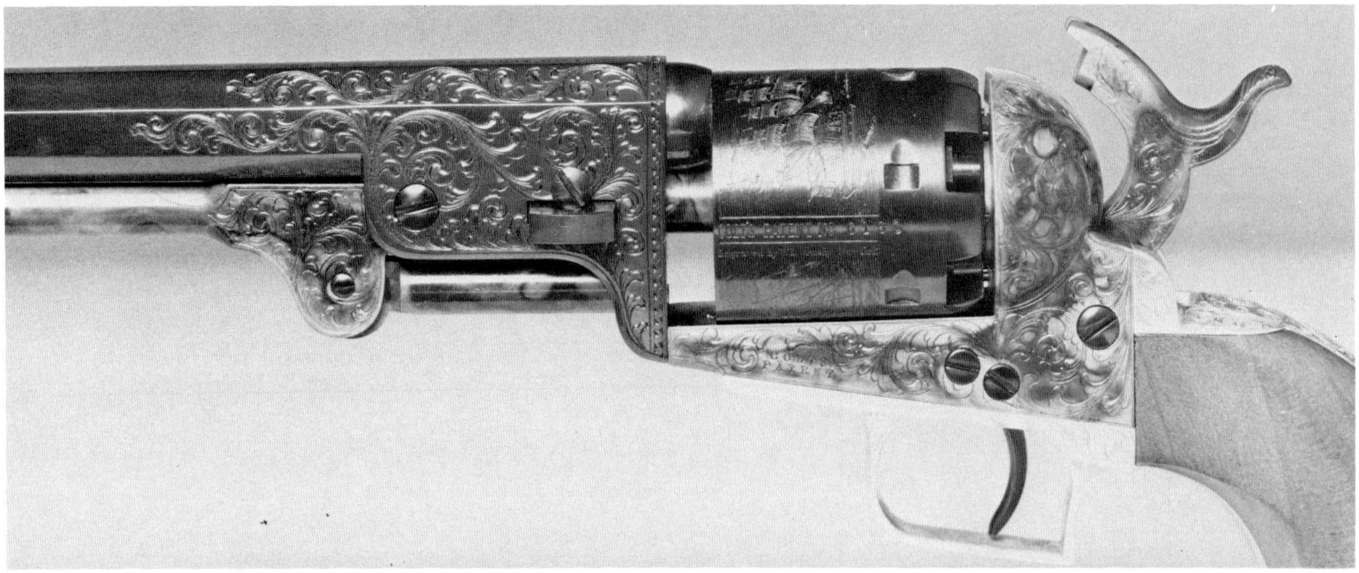

A modern Colt Navy revolver engraved by A. A. White Engravers, Inc. The early vine style scroll of 1848-51 is on the barrel, loading lever, frame, hammer (wolf head), and gripstraps.

Slide scene detail from the Chateau Thierry Custom grade World War I pistol. (Courtesy of Colt's Firearms Division)

"The Colonel Samuel Colt Special," another of the five deluxe Colt firearms engraved for Colt former president William H. Goldbach. This Single Action, with its extra cylinder and gold-plated cast sterling silver grips, is the prototype for the Centennial Model commemorative. The gun's right side features the evolution of the Colt theme in gold; the left hightlights the life of Colonel Colt. The gun abounds in Colt historical inscriptions, markings, and motifs covering the Single Action model's one hundred years of popularity and use. (Collection of William H. Goldbach)

**Two prototypes for possible commemoratives. The Single Action (right) was used as a reference for two guns Colt was considering: a hand-engraved Theodore Roosevelt issue and a similar model of the Golden Spike. The .45 automatic (left) features a European theater engraved decor on the left side and a Pacific theater theme on the right. It was designed as a possible hand-engraved deluxe pistol for the World War II commemorative series. Both guns are the work of Alvin White and are so signed.** (Collection of Joseph G. Berlin)

**A World War II theme is featured on the first of five deluxe Colt firearms done by Mr. White for Colt former president William H. Goldbach. The Government model automatic is silver- and gold-inlaid. Its left side commemorates Goldbach's Navy air experience; the right side, his Army air corps service. Named the "20th Air Force Commemorative," the gun displays Goldbach's name and title in gold inlay on the grip frame.** (Collection of William H. Goldbach)

# Mel Wood

*Santa Maria, California*

A graduate electronic and mechanical engineer, Mel Wood began teaching himself the art of engraving about 1970. While working for a large aerospace firm at Vandenberg Air Force Base, California, Mr. Wood began by engraving small steel plates to build up his skill before working on a firearm. He filled up two or three barrels full of practice plates. About four years later he engraved his first firearm, a Harrington and Richardson revolver that he still owns.

For his cutting, Mr. Wood uses both hammer and chisel and the Gravermeister, in conjunction with a heavy ball vise mounted on a rubber tire. In his designs he prefers to use his own distinct style of American scroll or oak-leaf decoration. Along with firearms he engraves quite a few handmade folding knives as well as an occasional belt buckle or jewelry.

In about 1975, when this author became interested in engraving while living in Santa Maria, California, Mel Wood helped me to learn the basics until I returned to Cincinnati to continue my apprenticeship. For his help and consideration during those beginning days, I am most grateful.

Since 1978 Mel Wood has been a full-time engraver and often displays his work at the various gun collector shows in California and the Knifemakers Guild show in Kansas City. He is a life member of the National Rifle Association and the Santa Maria Gun Club, as well as an officer in Elks Lodge #1538.

**Interlocking scrollwork dominates the design on this stainless steel Smith & Wesson. Note the use of a double running leaf pattern on the topstrap and triple running leaves on the cylinder. These motifs are used often on revolvers by Mel Wood.**

**A unique treatment is seen on this handmade folding knife. Mr. Wood has centered his design on the bolster and worked it outward in running vine fashion onto the blade and scales. Note the signature, MEL, on the blade and bolster.**

# Dwain R. Wright

*Bend, Oregon*

Born in 1946 and raised on a citrus ranch in the San Fernando Valley of Southern California, Dwain Wright has had a lifelong interest in art, arms, and the outdoors. His formal education began at Pierce State College, where he received an associate degree in wildlife management. Then he went on to Woodbury University where he received a bachelor's degree in commercial art and illustration. He eventually attained a master's degree in graphic design at the University of California at Los Angeles where he earned credits toward a PhD.

After college Mr. Wright worked as an art director and illustrator in the advertising business until he accepted work as a professor of graphic communications at the Linn-Benton Community College at Albany, Oregon. In this capacity he taught twenty-two different subjects in the fields of arts and crafts.

Mr. Wright is self-taught as an engraver and learned the art at the urging of Harold and Vernetta McCallum, who are collectors of Winchester arms. He attributes much of his knowledge and skill to the study of the 2,000 books in his personal library and the engraved European sporting arms in his collection, as well as the collections of others.

Mr. Wright works in all styles normally found on firearms, including inlays and overlays. He uses the hammer and chisel as well as the burin for all cutting, which he executes from both the engraver's block (ball vise) and the bench vise.

At present Mr. Wright is limiting his work to assignments of an intricate nature that require a great deal of time in the design and execution. As a result of this approach he can complete only two or three commissions per year.

**The left side of the C.I.A. Colt shows Mr. Wright's scrollwork.** (Collection of Amado Hernandez)

A Browning over/under shotgun, scroll engraved with flush gold borders and bank note style game scenes by Dwain Wright.

The underside of this Perazzi shotgun is a good illustration of Mr. Wright's skill at scrollwork. Note the fine line game scene.

This Perazzi Mirage engraved by Dwain R. Wright is 24 karat gold, inlaid in full relief. All borders are 24 karat gold wire. The bottom of the receiver is bulino. All engraving is in relief and the finish is French gray.

# Glossary

## A

ACANTHUS — A decorative leaf design used in a variety of ways by arms and bank note engravers. A typical example can be found between the upper and lower 20s on the back of a twenty dollar bill.

ALTDEUTSCHER LAUBGRAVUR — German term meaning old German leaf engraving. Austrian engravers use this term to describe what American engravers call Germanic floral engraving.

ALTENGLISCHE ARABESKEN — German term meaning old English scrollwork. Austrian engravers use this term to describe what American engravers call English scrollwork.

## B

BACKGROUND — The area that surrounds scroll or floral work. Background is usually matted with a punch in bas-relief work and cut away with a chisel in medium or high relief work.

BANK NOTE STYLE ENGRAVING — A method of engraving a figure or scene into metal with very fine lines so that the subject looks almost like a photograph to the naked eye. Example: the presidential portraits on United States currency are printed from bank note-engraved metal plates.

BAT WING — A decorative motif used as a border design, in the style of arms engraving made popular in America from 1860 to 1900.

BLOCK (Engraver's) — A special vise used by engravers. The block is usually spherical, with adjustable jaws on top, and sits in a felt, leather, or rubber ring.

BRIGHT CUT — A style of engraving commonly used on jewelry and western riding gear (spurs, bits, etc.). This type of engraving is also found on some knives and firearms. It is characterized by cuts that appear to be deep and reflective.

BULINO — The Italian word for a burin (hand graver). The term "bulino engraving" is used by some to describe a style of engraving scenes and figures popularized by certain contemporary Italian arms engravers. This style is known in the United States as bank note style engraving.

BURIN — A hand graver. A small engraving chisel with a mushroom-shaped wooden handle meant to be held and powered by the hand. Usually used for engraving soft precious metals or fine shading and details on harder metals, such as those used on arms.

BURNISHER — A polished steel hand tool used by engravers for rubbing down transfer patterns or removing light scratches from the metal being engraved.

# C

CARVE — As used by engravers, this term means to decorate by cutting or sculpting any material other than metal, such as wood, ivory, or stone.

CHASE — The ornamentation of metal by indenting with a hammer or another tool without a cutting edge. Also known as repousse.

CHASING — Engraving with hammer and chisel. This term is used because the type of tool employed by engravers is called a chasing hammer.

CHASING HAMMER — The hammer used by engravers (of steel or other hard metals) to strike the engraving chisel. The head of the chasing hammer usually has one circular end about 1 inch in diameter, and a peening or chasing protrusion on the other end. The handle is usually 5 to 8 inches long and is slim in its middle and bulbous on the end.

CHISEL — Sometimes called the long chisel, this is the arms engraver's basic tool for executing his designs. The length of the chisel varies from one engraver to another, but is usually between the burin and the die sinker's chisel in length. It can be made with one of several point shapes, including onglette, round, flat, knife, square, lozenge, bevel, and oval. Most firearms engravers use a tough grade of high speed steel for their chisels because of the extremely hard metal found in modern firearms. Other engravers can usually work with tool steel chisels because of the softer metals that they engrave.

CIPHER — A monogram. The intertwining of two or more letters to form a unique design. Presentation arms are often engraved with the owner's ciphered initials.

# D

DAMASCENE GOLD WORK — A process of arms decoration. First the subject is plated or overlaid with gold; then the design (usually Arabic in style) is cut through the gold, exposing the metal underneath; finally the piece is subjected to a bluing agent that leaves the exposed metal dark blue or black and the gold untouched. At present this type of decoration is done almost exclusively in Spain.

DIE SINKER'S CHISEL — An engraving chisel used by steel die engravers. Much longer than the usual arms engraver's chisel, it usually measures about 7½ inches. Some arms engravers have a preference for this tool.

DRAGON HEAD HAMMER — The hammer of a firearm (usually a revolver) engraved to look from the side like the head of a dragon. Variations include an eagle and wolf head. At one time percussion revolvers engraved with the wolf motif were thought to be the work of a jewelry engraver, Joseph Wolf, but were later found to be the work of Louis D. Nimschke of New York, who was active from 1850 to 1900.

# E

ENGRAVE — As used by arms engravers, this term means the cutting of a design or motif into metal with a chisel or burin. It is not used to describe the decoration on nonmetal objects such as wood, ivory, stone, or horn. These items are carved, scrimshawed, or sculptured. Note: Firearms and edge weapons decorated with mass production methods such as roll stamping, acid etching, or photo etching are often incorrectly referred to as engraved. In some cases this improper use of arms terminology is caused by lack of knowledge, but sometimes it is a deliberate attempt to deceive the potential purchaser. Since hand-engraved pieces are almost always more valuable than the same items decorated by mass production methods, the buyer should be certain of what he is ordering. Though some reputable arms engravers do one-of-a-kind etchings, they never represent etching as engraving.

ENTWINING LINE MOTIF — A design, either engraved or inlaid, which uses two or more lines that alternately separate and converge in a uniform manner.

ETCH — To produce a design, especially on metal or glass, by the corrosive action of an acid. In arms embellishment, etching is sometimes properly used in conjunction with engraving in order to achieve a special effect on the background or finish of the metal. This practice is entirely acceptable and even desirable in some cases. Most sophisticated firearms collectors, however, avoid those pieces entirely decorated by etching from set patterns, because their decoration is just another form of mass production.

# F

FESTOON — A motif representing a decorative chain of foliage or strip hanging between two points.

FIGURE — Engravers use this term to describe a single subject to be engraved or inlaid, such as a human being, an animal, or a mythological creature.

FINISH — The final treatment given to the metal parts of a

firearm after engraving. Some finishes are: bluing; browning; plating (including nickel, silver, or gold); graying (including French gray, satin gray, frost gray, and acid gray); and Parkerizing. Unlike the jewelry or die engraver, the arms engraver must be familiar with all finishes and must be able either to apply them himself or to subcontract this work to a qualified gunsmith.

FISH FIN — A decorative motif, usually used for border decoration, made popular by nineteenth century arms engravers, but used much earlier.

FLORAL — A style of arms decoration that uses more flowers and leaves than scrollwork, which depicts spiral vines. A style known as floral-scroll uses spiral vines in conjunction with flowers and leaves.

FLORENTINE — 1. Used in the decoration of jewelry, the term means to apply a rough, shaded texture to the piece with a burin. 2. A type of textured background surrounding the vines and leaves of scroll or floral designs. 3. Another term for floral engraving.

FLYOUT — An engraver's term to denote a slip with the burin or chisel.

FRET WORK — A design of usually angular interlocking lines, originating in ancient Greece, it is used by engravers for cut or inlaid borders.

## G

GRAVER — A general term for any tool used to engrave metal.

GRAVERMEISTER — The trade name for a pneumatic-powered engraving chisel. The Gravermeister was developed by the contemporary engravers John Rohner and Donald Glaser and uses a regular chisel set in a handpiece and connected to a length of pneumatic tubing. The tubing is connected to a motor that provides an alternating pulse of air, which activates a hammer within the handpiece. Though the Gravermeister is a mechanical device, work done with it is considered hand engraving because the pneumatic hammer only takes the place of the chasing hammer, freeing one hand to turn the vise; cutting done with it is only as good as its operator.

GRIFFIN — A mythical creature, part eagle and part lion, sometimes used as an engraved motif, especially in earlier centuries.

GROTESQUE — A decorative motif characterized by fanciful or fantastic human and animal forms, often interwoven with foliage or similar figures; the natural may be distorted into absurdity, ugliness or caricature. Grotesques were especially popular with arms embellishers from the seventeenth through the nineteenth centuries. Few of America's modern arms engravers venture into this baroque style of decoration, though the effect of such decoration can be splendid.

## I

INLAY — In arms decoration an inlay is precious metal, such as gold, silver, or platinum, recessed into steel; a dovetail undercut is used to prevent the precious metal from coming out. Inlays fall into the following four general categories: 1. *Raised figure* — the inlay is slightly higher than the steel into which it is inlaid, and is usually modeled or sculpted. 2. *Flush figure* — the inlay is flush with the steel and is usually detailed like a bank note or line engraving. 3. *Raised line inlay* — the inlaying of precious metal wire to form border lines or spirals; the wire is left slightly higher than the surrounding steel. Raised lines are rarely used except on exhibition pieces, because they are more easily damaged or worn down. 4. *Flush line inlay* — the same as raised line inlay except that the lines are finished flush with the steel.

## L

LINE ENGRAVING — 1. A scene or figure engraved into metal using fine lines, not so detailed or realistic in appearance as a bank note style engraving. 2. Scroll or floral engraving without background matting or relief.

LINER — An engraving chisel that cuts several parallel lines simultaneously.

LOUPE — A small magnifying glass made to be held in the recess of the eye, used by jewelers and engravers.

## M

MOTIF — A decorative design, such as an eagle and shield, police badge, griffin, arrow, grotesque, and sunburst.

## N

NGRAVER — A hand-held, electric-powered engraving device. Similar in concept to the Gravermeister, but powered by an electric motor via a flex shaft.

NEUENGLISCHE ARABESKEN — German term meaning new English scrollwork. Austrian engravers use this term to describe what American engravers usually call Germanic scrollwork.

## O

OAK LEAF — A design, commonly engraved on Austrian and German arms, consisting of oak leaves and acorns.

OVERLAY — The process of attaching a precious metal design or figure to the surface of the arm to be decorated. This can be done by various methods, including soldering, brazing, plating, or roughing the surface and hammering on the overlay. Most arms engravers avoid overlays, both because they can not be made flush with the steel and because they are much more likely, than inlays, to fall off.

## P

PHOTOENGRAVING — A photomechanical process for making line cuts and halftone cuts by photographing an image on metal and then etching. For firearms, this process is used only on a mass production basis.

PULL — A print made by one of several methods from the engraved surfaces of a firearm. Some pulls appear in negative, by printing from the top surface, and some in positive by drawing the ink or lampblack out of the cuts. These pulls are saved by the engraver as a record of his work.

PUNCH — A tool meant to be struck by a hammer so as to imprint a design, letter, or numeral into metal. Punches are most often used by arms engravers to matte the background of scroll or floral designs. Backgrounding punches usually impress a tiny circle or dot into the metal.

PUNCH WORK — 1. To American engravers this term means the decoration of a firearm entirely with a punch or punches. The best-known examples of this type of work are found on Merwin and Hulbert firearms of the late nineteenth and early twentieth centuries. 2. To British engravers this term means an inexpensive grade of factory engraving done by two engravers. One cuts the spirals with a hammer and chisel; the other cuts the inner leaves with a burin.

## R

RELIEF ENGRAVING — Cutting away of the background metal so that the designs or figures appear to project above the surface of the metal. High relief engraving has the background cut away deeper than medium relief. Bas-relief, also known as low relief, does not have the background cut away; the background is stippled or matted instead.

RIFFLER — A small fine file, made in various shapes, used to sculpt or model relief engraving or inlays.

ROSETTE — A circular and symmetrical design used by engravers to decorate circular projections on firearms and knives, such as hinge pins and rivets in knife handles.

## S

SCENE — A design incorporating more than one figure, sometimes having a landscape or building in the background and/or foreground.

SCRAPER — A hand tool used by engravers to scrape small amounts of metal while modeling or sculpting a design.

SCRIBER — A tool used by engravers and machinists to scratch a mark on a hardened surface. The scriber resembles a steel pencil and has a needle-like point. Also known as a scratch awl.

SCRIMSHANDER — One who scrimshaws.

SCRIMSHAW — Scratching or cutting a shallow design into horn, bone, or ivory, done in a style that resembles line or bank note style engraving on metal. Many fine arms have scrimshawed grips or accessories.

SCROLL — A spiral form in ornamental design derived from the curves of a loosely rolled parchment scroll. This is the most widely used and varied type of decoration traditionally engraved into firearms. The scrollwork used on arms falls into several design categories. The name for each type varies from one engraver to another, because there is no guild or formal school for engravers in the United States to standardize such terms. The following is a list of commonly used terms: 1. *German scroll* — medium to small spirals that wind up like watch springs. Inside the spiral are tiny leaves which are separated by background matting or relief. 2. *Black leaf* — to most engravers, the same as German scroll. 3. *American scroll* — developed by nineteenth century engravers who came to America from Europe, it is usually composed of medium to large spirals with fewer leaves than German scroll and less windup. The innermost part of the spiral usually has a large lobe or cup. 4. *Victorian scroll* — similar to American scroll but more finely shaded and with different types of leaves. 5. *English scroll* — small to very small spirals in which there are uniform leaves not separated by background. 6. *Arabesque* — most American engravers use this word to mean English scroll; some Austrian-trained engravers use it to describe a very leafy type of floral scroll; the dictionary and yet other engravers understand the word to mean an Arabic-type design, such as that used in Persian rugs or gold damascene work.

STONE — A rectangular stone used by engravers to sharpen their gravers.

## T

TRANSFER — One of several types of patterns used by

engravers to transfer a predetermined pattern to the metal to be engraved.

TRANSFER WAX — A special wax-like substance used to reproduce the cut lines of an engraving.

## W

WHEAT CHAFF — A design used for borders, it resembles wheat chaff laid end-to-end.

WHITE — Sometimes called Chinese White, a type of paint used to cover the metal before engraving so that the design can be sketched or transferred on.

WIGGLE TOOL (Wriggle Tool) — A burin shaped like a flat chisel, used to engrave zigzag lines.

## Z

ZIGZAG — The design cut by a wiggle tool. Usually used as decoration on western revolvers as well as western riding gear and belt buckles.

# Directory of Engravers

## A

ACEVEDO, HORACIO
Hieland Road
Kankakee, Illinois 60901

ADAMS, JOHN
47 Brown Avenue
Mansfield, Massachusetts 02048

ALPEN, RALPH
Box 7, Bentely Road
West Grove, Pennsylvania 19390

AVERILL, ED
60 Chesnut Street
Cooperstown, New York 13326

## B

BARBETTI, PAOLO
2218 5th Avenue
Seattle, Washington 98121
*Brochure with prices — free.*

BAYER, JOSEPH C.
Sunset Avenue, Sunset Hill,
Rural Delivery 1
Princeton, New Jersey 08540

BEE, ANGELO
10703 Irondale Avenue
Chatsworth, California 91311

BEEBE, BRUCE
124 East Avenue
Akron, New York 14001

BLEDSOE, WELDON
6812 Park Place Drive
Ft. Worth, Texas 76118

BLEILE, CARL B. (BOYD)
Post Office Box 11285
Cincinnati, Ohio 45211
*Brochure — $5.*

BLEILE, C. ROGER
Post Office Box 5112
Cincinnati, Ohio 45205

BONHAM, HENRY D. (HANK)
218 Franklin Avenue
Seaside Heights, New Jersey 08751

BRIDGES, B. BRYAN
6350 East Paseo San Andres
Tucson, Arizona 85710

BROWN, HOWARD E.
Rural Route 4, Box 91, Shaved Acres
Winchester, Indiana 47394

BURT, BOB
25 Alto Road
Burlington, Connecticut 06013

# C

CARGNEL, GINO
6466 East Sibley
Simi Valley, California 93063

CHURCHILL, WINSTON G.
Twenty Mile Stream Road,
Route 1, Box 29 B
Proctorsville, Vermont 05153

CROCKER ENGRAVING
510 42nd Street
Los Alamos, New Mexico 87544

# D

DARAKIS, ART A.
Rural Delivery 1, Levitt Road
Amhurst, Ohio 44001

DAVIS, JIM
230 South Main Street
Eldorado, Ohio 45321

DE MUNCK, JAMES R.
3012 English Road
Rochester, New York 14616

DOVE, HOWARD M.
52 Brook Road
Enfield, Connecticut 06082

DUBBER, MICHAEL
3107 East Mulberry
Evansville, Indiana 47714
*Price list — $1.*

# E

ENGLISH, WILTON L.
12009 Barksdale Drive
Omaha, Nebraska 68123

EVANS, ROBERT (BOB)
332 Vine Street
Oregon City, Oregon 97045

# F

FAVRE, JAQUELINE
3212-B Wynn Road, Suite 214
Las Vegas, Nevada 89012

FINLEY, CLINTON
4176 St. Patrick Avenue
Redding, California 96001

FLANNERY, JEFFREY W.
Gunpowder Creek Gunworks,
Route 1, Box 155
Union, Kentucky 41091

FLIGER, LYNN
5036 Hughes Avenue, NE
Fridley, Minnesota 55421

FORTE, GUISEPPI
1126 West Jeffers Drive
Rawlins, Wyoming 82301

FRANCOLINI, LEONARD
Post Office Box 32
West Granby, Connecticut 06090

FRANK, HEINRICH H. (HENRY)
Route 1, Mountain Meadows
Whitefish, Montana 59937

FRENCH, J. RON
2633 Quail Valley
Irving, Texas 75060

# G

GILES, ED F.
204 Fremont Street
Rehoboth, Massachusetts 02769

GLASER, DONALD
1520 West Street
Emporia, Kansas 66801

GOLD, ERIC
Post Office Box 1904
Flagstaff, Arizona 86002

GOODWIN, DANIEL
Post Office Box 1619
Kalispell, Montana 59901

GRANT, HOWARD V.
Post Office Box 396
Lac du Flambeau, Wisconsin 54538

GURNEY, F. R.
2301-995 Jasper Avenue
Edmonton, Alberta, Canada T5J 2X4

GWINNELL, BRYSON J.
26 Fairview Avenue
Marlboro, Connecticut 06447

# H

HAAS, THOMAS SR.
Rural Route 2
Spencer, Indiana 47466

HARRINGTON, FRED A.
2107 West Frances Road
Mount Morris, Michigan 48458

HARTLIEP, GLEN
6130 East 30th
Tucson, Arizona 85711

HARTLIEP, NEIL
(*December-June*)
Post Office Box 50805
Tucson, Arizona 85703
(*July-November*)
Post Office Box 733
Fairmont, Minnesota 56031

HENDRICKS, FRANK E.
Route 2, Box 189 J
San Antonio, Texas 78229

HEUNE, BENNO L.
934 Jack London Drive
Santa Rosa, California 95405

HICKS, TOM
4316 31st Street
Lubbock, Texas 79410

HIPTMAYER, HEIDEMARIE
1771 Harvey Avenue
Kelowna, British Columbia,
Canada V1Y 6G4

HOPEWELL, L.
2062 Truesdale Lane
Carlsbad, California 92008

HUFF, STEPHEN E.
Post Office Box 8663
Missoula, Montana 59807

HURST, KEN
Room 207, Krise Building, 9th Street
Lynchburg, Virginia 25404

## I

INGLE, RALPH W.
4 Missing Link
Rossville, Georgia 30741
*Catalog and price list — $2.*

## J

JOHNS, BILL
1113 Nightingale
McAllen, Texas 78501

## K

KAIN, ROBERT C. (BOB)
Post Office Box 66
Canaan, Connecticut 06018

KAMYK, STEVEN
50 Church Street
West Springfield,
Massachusetts 01089

KAYE, TOMMY J. (T. J.)
4745 Dellwood
Beaumont, Texas 77706

KELLY, LANCE
4226 Lamar Street
Decatur, Georgia 30035

KELSO, JIM
Post Office Box 114
Baring, Washington 98224

KOEVENIG, E. J.
Keystone, South Dakota 57751
*Free brochure.*

KOLOUCH, WALTER
Route 2, Box 289
McMinnville, Oregon 97128

KUDLAS, JOHN M.
622 14th Street SE
Rochester, Minnesota 55901

KUSMIT, JOHN
35 King Street
Wallingford, Connecticut 06492

## L

LAGEOSE'S ENGRAVING
624 LaSalle Road
Kalispell, Montana 59901

LANE, BEN JR.
2118 Lipscomb Street
Amarillo, Texas 79509

LANE, BETH
201 South Main Street
Pontiac, Illinois 61764

LEWIS, W. NEAL
6300 Mixon Road
Palmetto, Georgia 30268

LINDSAY, FRANK
1326 Tenth Avenue
Holdrege, Nebraska 68949

## M

MACHU, EDWARD JOE JR.
5306 Broadway
San Antonio, Texas 78209

MAINS, WILLIAM H. (BILL)
3212B Wynn Road, Suite 214
Las Vegas, Nevada 89102
*Catalog and prices — $1.*

MAKI, ROBERT
814 Revere Road
Glenview, Illinois 60025

MAREK, RUDY
Route 1, Box 1A
Banks, Oregon 97106

MARKTL, FRANZ
Post Office Box 716
Kalispell, Montana 59901

MC KENZIE, LYNTON S. M.
6940 North Alvernon Way
Tucson, Arizona 85718

MILLER, MICHAEL JON
8734 Lukens Lane
Alexandria, Virginia 22309

## O

OFFENBERGER, WARREN E.
Star Route
Reno, Ohio 45773

OKER, KONRAD SR.
280 Illinois Street
Crystal Lake, Illinois 60014
*Free brochure available.*

## P

PEDINI, MARCELLO
470 Deer Park Avenue
Dix Hills, New York 11746

PFEIFFER, HANS
286 Illinois Street
Elmhurst, Illinois 60126

PIERCE, BARBARA R.
248 East Ridgeway
Hermiston, Oregon 97838

PIQUETTE, PAUL R.
40 Royalton Street
Chicopee, Massachusetts 01020

PITETTI, ARTHUR
Hawk Hollow Road
Denver, New York 12421

POTTS, JEREMY
Post Office Box 85
Pine Bluff, Wyoming 82082

POTTS, WAYNE E.
912 Poplar Street
Denver, Colorado 80220

PRUDHOMME, EDWARD C.
(JACK)
513 Ricou-Brewster Building
Shreveport, Louisiana 71101

## R

RABENO, MARTIN
Post Office Box 37 F. Rd. 1
Ellenville, New York 23428

ROHNER, JOHN and HANS
Sunshine Canyon
Boulder, Colorado 80302

RUNDELL, JOE
6198 Frances Road
Clio, Michigan 48420

## S

SCOTT, ALVIN E.
2610 Massey Tompkins
Baytown, Texas 77521
*Price list available.*

SHERWOOD, GEORGE
P.O. Box 735
Winchester, Oregon 97495
*Free catalog and price list.*

SHOSTLE, BEN
1201 Burlington Drive
Muncie, Indiana 47302
*Free brochure and prices.*

SINCLAIR, BURT
Post Office Box 541
Shallowater, Texas 79363

SKAGGS, RON
508 West Central
Princeton, Illinois 61536
*Brochure available.*

SMITH, RON
5408 Bowie
Fort Worth, Texas 76107

SMITH, RUSS J.
231 Springdale Road
Westfield, Massachusetts 01085

SPRING, GEORGE B. III
Rural Free Delivery 4, Route 82
Salem, Connecticut 06415

SWARTLEY, ROBERT
2800 Pine Street
Napa, California 94559

SWARTZFEGER, MONK
470 Duncan Avenue
Washington, Pennsylvania 15301

# T

TUSCANO, ANTHONY (TONY)
1473 Felton Road
South Euclid, Ohio 44121

# V

VALADE, ROBERT B.
Route 1, Box 30A
Cove, Oregon 97824

VEST, JOHN
6715 Shasta Way
Klamath Falls, Oregon 97601
*Price list — $1.*

VIRAMONTEZ, RAYMUNDO (RAY)
4348 Newberry Court
Dayton, Ohio 45432
*Catalog and prices — $3.*

# W

WAGONER, VERNON G.
2340 East Fox
Mesa, Arizona 85203

WALLACE, TERRILL LEE (TERRY)
385 San Marino
Vallejo, California 94590

WARREN, FLOYD E.
1273 Wilson Sharpsville Road, NE
Cortland, Ohio 44410

WARREN, JOHN E.
Post Office Box 72
Eastham, Massachusetts 02642

WELCH, GEORGE S. (SAM)
Post Office Box 2152
Kodiak, Alaska 99615

WELLS, RACHEL
110 North Summit Street
Prescott, Arizona 86301

WHITE, ALVIN A.
Post Office Box 68
Manchester, Connecticut 06040

WOOD, MEL
3901 Crestmont Drive
Santa Maria, California 93454

WRIGHT, DWAIN R.
67168 Central
Bend, Oregon 97701